70 Times 7 Math:
Tests

(Kindergarten Through 5th Grade)

Published by 70 Times 7 Math (A division of Habakkuk Educational Materials)

Copyright © 2019-2023 by 70 Times 7 Math. All rights reserved.

70 TIMES 7 MATH: TESTS
(KINDERGARTEN THROUGH 5ᵀᴴ GRADE)

Copyright © 2019-2023 by 70 Times 7 Math

All rights reserved. No part of this book may be reproduced in any form or by any electronic or mechanical means, including information storage and retrieval systems, without the written consent of the publisher. Please address your inquiries to Habakkuk@cox.net.

ISBN (Paperback Edition): 978-1-954796-29-4

Image on the front cover and title page: © [Deezaat] / Adobe Stock

Geometric 3d-Line Shapes: © YummyBuum/Shutterstock.com

Printed and bound in the United States of America

Published by 70 Times 7 Math
(A division of Habakkuk Educational Materials)

Visit www.habakkuk.net

This test book supplements the kindergarten through 5th-grade textbook *70 Times 7 Math (An All-In-One Math Book for Grades Kindergarten Through 5th)*. Besides the textbook, students will also need the *70 Times 7 Math: Classwork/Homework (Kindergarten Through 5th Grade)* book to help prepare them for the exams. These can be purchased from the Habakkuk Educational Materials website. (**Computer-based tests** and **practice tests** are also available through the website.) The classwork/homework assignments are meant to provide students individual practice solving problems taught in their textbooks and to equip them to do well on the corresponding tests. Questions included on the classwork and tests are not randomly selected from the chapters in focus. Instead, they are very comprehensive in that every type of problem students learn about in their textbook and complete as a class on the interactive whiteboard will also be solved individually by students on the classwork assignments and tests. The answer keys to the tests are provided at the end of the book.

The number of tests administered yearly depends on the grade level of the students. There will be 4 to 6 comprehensive tests administered yearly to elementary school students. Four tests are available for students in grades kindergarten through 2nd, while students in grades 3rd through 5th have six tests to complete. Since students are tested over every type of problem included in their textbook, some of these comprehensive tests may take more than one class period to complete if you have a fixed amount of time for math each day. Please allow every student to finish each test without penalty. To contact Habakkuk Educational Materials, please visit the business website at the address below.

https://www.habakkuk.net/

Class Supplies List for Students

1. Three-ring binder, notebook paper, zipper pencil bag
2. Pencils, erasers, and pencil sharpener
3. Ruler
4. Protractor for grades 3rd-5th
5. Calculator for grades 3rd-5th (Recommended: CASIO *fx-9750GII*)
6. *Polygons, Polyhedrons, and Other Shapes for Grades Pre-K Through 5th*, by 70 Times 7 Math (**Note:** It is optional but not necessary that each student have a copy of this book. It is recommended that every kindergarten through 3rd-grade teacher have the eBook version to use as digital flashcards throughout the year to teach students the shapes and to prepare them for their last test.)
7. Textbook--
   70 Times 7 Math: An All-In-One Math Book for Grades Kindergarten Through 5th
8. Classwork/homework book--
   70 Times 7 Math: Classwork/Homework (Kindergarten through 5th Grade)
9. Test book--
   70 Times 7 Math: Tests (Kindergarten Through 5th Grade)
10. Access to the computer-based tests (See the next two pages for details.)

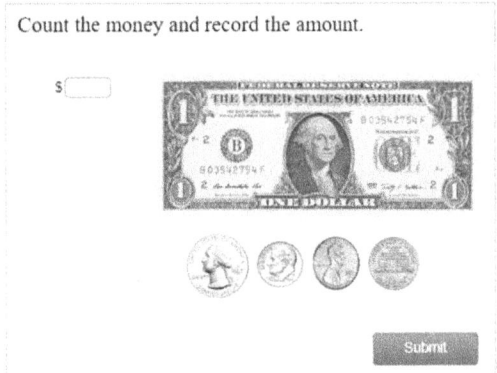
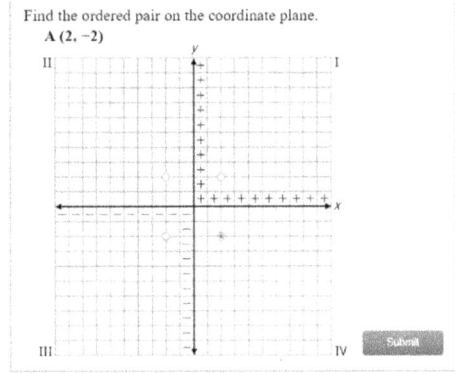

To purchase these materials, please visit the "Math Materials" page of the Habakkuk Educational Materials website at https://www.habakkuk.net/.

70 Times 7 Math: An All-In One Math Book for Grades Kindergarten Through 5th

70 Times 7 Math: Classwork/Homework (Kindergarten Through 5th Grade)

Polygons, Polyhedrons, and Other Shapes for Grades Pre-K Through 5th

COMPUTER-BASED TESTS FOR THE "70 TIMES 7 MATH CURRICULUM"

Overview of the online tests and how to gain access: Computer-based tests specially designed to assess a student's knowledge of the information covered in this book are available on Moodle, a Learning Management System. To gain access to the tests, please visit the business website of Habakkuk Educational Materials at https://www.habakkuk.net/ and click on "Math Materials" from the homepage. There you will find an online request form to fill out and a flyer to download with more information about these online tests. Habakkuk Educational Materials will finalize your enrollment within 48 hours of purchase (please refer to the form for pricing details), and you will have complete access to the online tests during the enrollment period, which includes both practice tests and a regular test for each chapter of the textbook. Please note that paper versions of the tests are available in *70 Times 7 Math: Classwork/Homework (Kindergarten Through 5th Grade)* and *70 Times 7 Math: Tests (Kindergarten Through 5th Grade)*.

How to use the tests: It is recommended that students originally be allowed to use their books to guide them through the practice tests. When the student feels confident or at the teacher's discretion, there is also a computer-based chapter test they can take to test their comprehension over the book's content. After submitting an answer to a question, the computer will notify test takers if their answer is correct or incorrect. When taking the practice tests, students are often given a second chance to answer the question correctly, and tutorial videos are available to provide them assistance as needed. After entering their answer to the final question of a test or practice test and clicking "Continue," a percentage grade will be available and a "Review Course" option to review any incorrect answers will also be accessible.

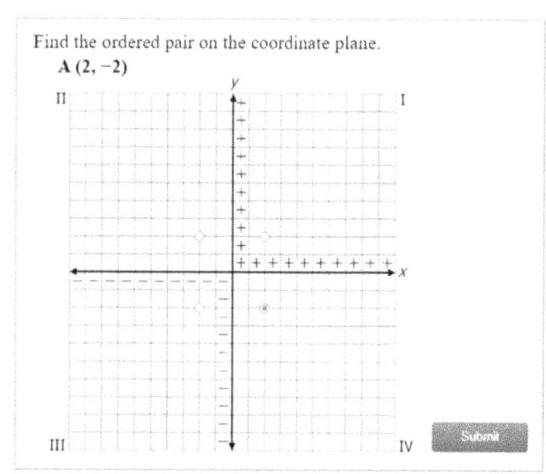

The tests serve two purposes:
(1) to assess a student's mastery of the content in this book; and
(2) to be used for review games.

Using the online practice tests for pretests and for review: The practice tests can be used as **pretests** at the beginning of a school year to determine what students already know, and parents could be given a summary of the results. The practice tests also provide an opportunity for students to independently review material covered on previous tests. Beginning the first week following their first test, students spend once a week at the computers answering questions over material that they have already been tested over so as not to forget it once the new information has been introduced.

Using the online tests for review games: The computer-based tests can be used in correspondence with various board games that have a pathway from start to finish. The review games could be played in math centers, or the entire class could be divided into small groups and a gameboard provided for each set of students. Children playing the game would answer one of the questions, and if the computer confirms that the answer is correct, the student could roll dice or spin a spinner and move his or her playing piece the corresponding number of spaces on the path.

If you visit the "Free Teaching Materials" page of the Habakkuk Educational Materials website, there are free gameboards that can be printed on cardstock. (Click on "Board games and more to complement Habakkuk Educational Materials' Bible, reading, language, math, science, and social studies materials.") The coupons referred to in the directions of one of the gameboards can be printed from the same file, and students who earn a "Good Work Coupon" might be designated a certain amount of time to play math games on the computer or whatever other privilege the teacher might choose.

Table of Contents

TESTS FOR GRADES KINDERGARTEN THROUGH 5TH

    Beginning Geometry Test (grades kindergarten through 2nd)   8

1. Elementary Test #1 (grades kindergarten through 5th)   10
2. Elementary Test #2 (grades kindergarten through 5th)   18
3. Elementary Test #3 (grades kindergarten through 5th)   31

TESTS FOR GRADES 3RD THROUGH 5TH

4. Elementary Test #4 (grades 3rd through 5th)   43
5. Elementary Test #5 (grades 3rd through 5th)   55
6. Elementary Geometry Test #6 (grades 3rd through 5th)   64

ANSWER KEYS

7. Key for Beginning Geometry Test (grades kindergarten through 2nd)   73
8. Key for Test #1 (grades kindergarten through 5th)   75
9. Key for Test #2 (grades kindergarten through 5th)   83
10. Key for Test #3 (grades kindergarten through 5th)   96
11. Key for Test #4 (grades 3rd through 5th)   108
12. Key for Test #5 (grades 3rd through 5th)   120
13. Key for Geometry Test #6 (grades 3rd through 5th)   129

Name: _____ Date: _____

Beginning Geometry Test
(Grades: kindergarten, 1st, 2nd)

Text: *Polygons, Polyhedrons, and Other Shapes for Grades Pre-K Through 5th*

Record the number of sides, angles, and vertices each polygon has:
   3, 4, 5, 6, 7, 8, 9, 10, or *unknown*.

1. nonagon   _____
2. octagon   _____
3. pentagon   _____
4. heptagon   _____
5. *n*-gon   _____

6. quadrilateral _____
7. decagon   _____
8. triangle   _____
9. hexagon   _____
10. What is a 3-gon? _____

Identify the number of faces each polyhedron has:
   4, 5, 6, 7, 8, or 10.

11. octahedron   _____
12. hexahedron   _____
13. pentahedron _____

14. decahedron   _____
15. tetrahedron _____
16. heptahedron _____

17. Cross out the shapes that are not polygons.

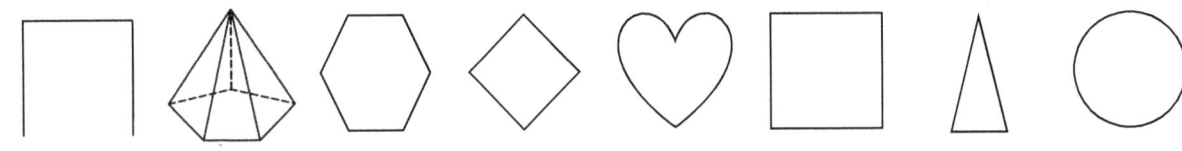

18. Circle all the quadrilaterals on the next page.

19. Which shape on the next page can be called a decagon? _____

Directions: Write the name of each flat shape in the space provided. The first one has been done for you.
  circle, cross, diamond/rhombus, heart, heptagon, hexagon, nonagon, octagon, oval, ~~parallelogram~~, pentagon, rectangle, square, star, trapezoid, triangle

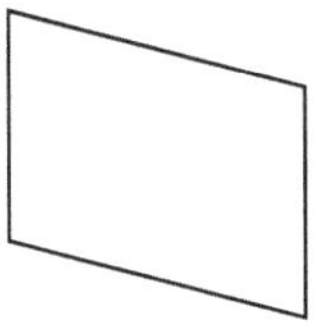

20. parallelogram 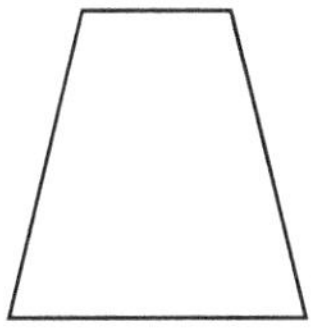   21. 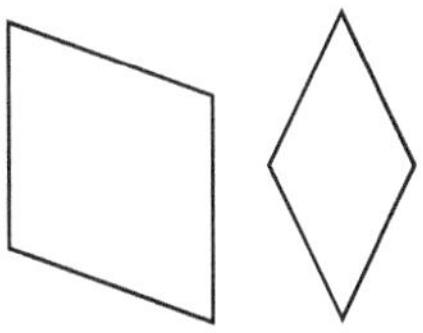 _____   22. 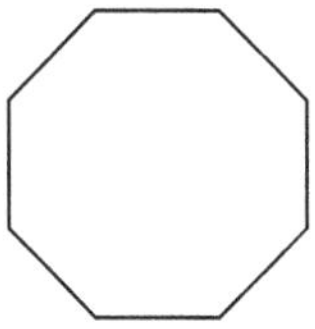 _____   23. _____

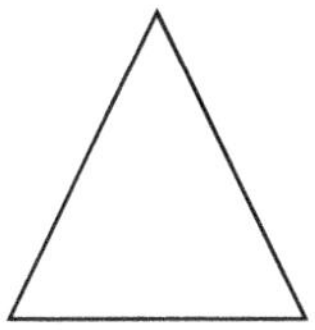

24. 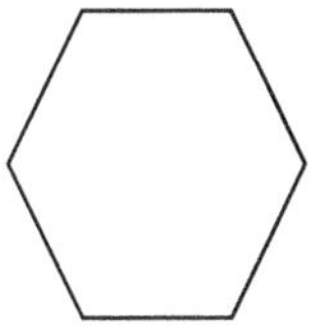 _____   25. 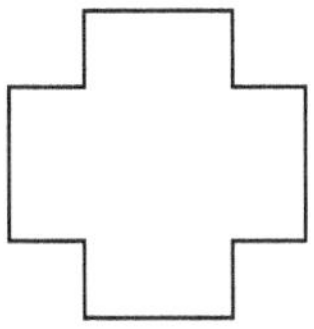 _____   26. 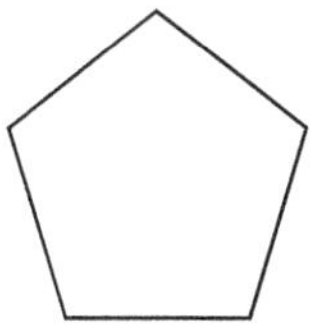 _____   27. _____

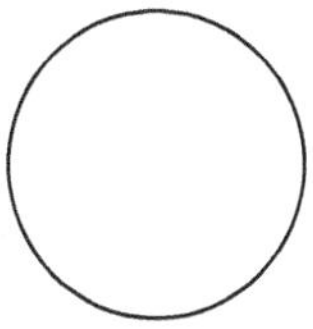

28. 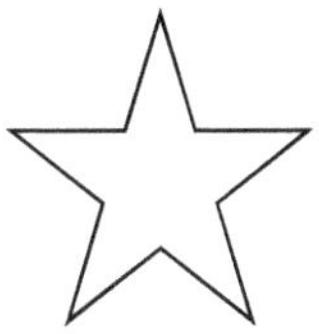 _____   29. 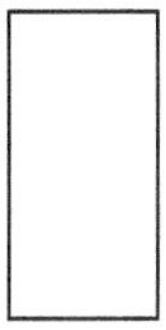 _____   30. 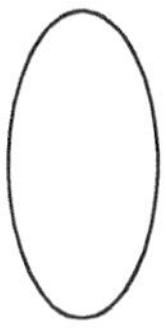 _____   31. _____

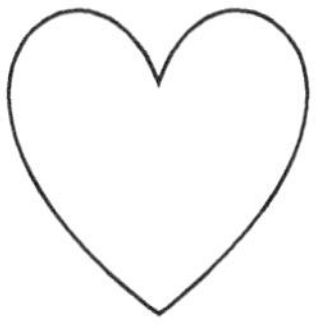

32. 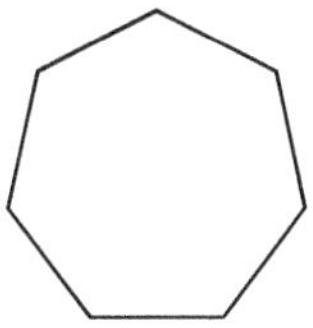 _____   33. 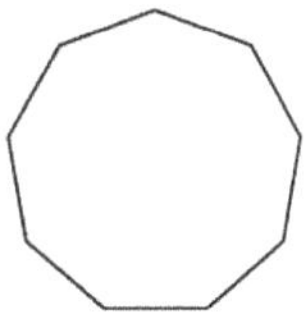 _____   34. _____   35. _____

Directions: Write the name of each solid figure in the space provided.
  cone, cube, cylinder, sphere

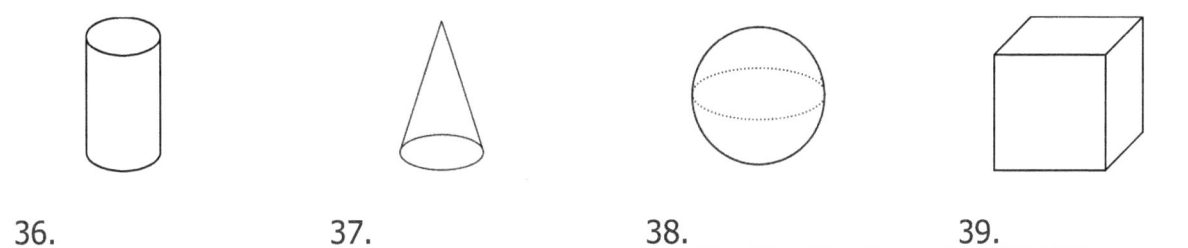

36. _____   37. _____   38. _____   39. _____

Name: _____ Date: _____

Elementary Math Test #1
(Grades: kindergarten, 1st, 2nd, 3rd, 4th, 5th)

ODD AND EVEN NUMBERS

Directions: Categorize the numbers listed below.

Numbers: −10, 2, 9, 15

    Odd   **Even**

1. _____   3. _____

2. _____   4. _____

5. Record any 2-digit <u>odd</u> number. _____

6. Record any 2-digit <u>even</u> number. _____

NUMBER VALUES

7. How many arrows are in the set? Write the **cardinal** number in the space provided.

    _____

8. What place is the underlined swimmer in? Write the **ordinal** number in the space provided.

    _____

ROMAN NUMERALS

9. List the first 12 Roman numerals in order from 1 to 12.

    _____

ADDITION AND SUBTRACTION FACTS

Directions: Solve.

10. 3 + 5 = _____   12. 13 + 0 = _____

11. 7 − 2 = _____   13. 9 + 1 + 4 = _____

Copyright © 2019-2022 by 70 Times 7 Math (a division of Habakkuk Educational Materials). All rights reserved.

Solve the equations. Show your work in the box.

14. $n - 7 = 9$ $n =$ _____

15. $n + 5 = 15$ $n =$ _____

Fill in the blanks. Use the words *difference, product, quotient,* or *sum*.

16. Addend + addend = _____

17. $n - n =$ _____

18. Factor × factor = _____

19. Dividend ÷ divisor = _____

Directions: Which pair of numbers equal the sum 8 when added together and the difference 4 when subtracted? Write the solution to the problem in the space provided.

20. Sum is 8
    Difference is 4

    _____ (6, 2   5, 3   7, 1)

PLACE VALUE

21. _____ hundreds   _____ tens   _____ ones

    Total number: _____

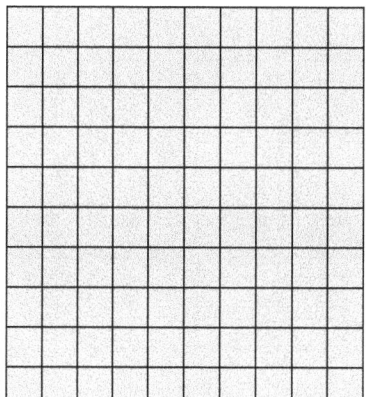 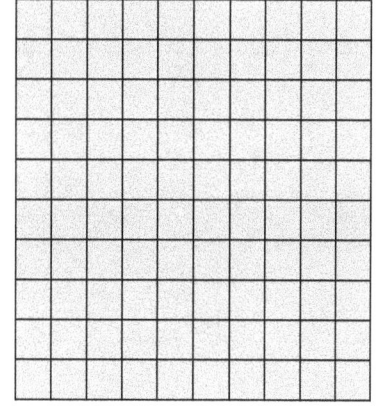 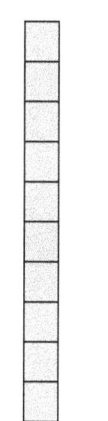

22. In the number 128, identify the number in the ones place, tens place, and hundreds place.

Ones	Tens	Hundreds

23. Fill in the blanks.

_____ hundreds + _____ tens + _____ ones = 682

Directions: Write the value of each underlined digit in the space provided.

<u>7</u>92   7<u>9</u>2   79<u>2</u>

24. _____   25. _____   26. _____

27. In the spaces provided, rewrite each number in standard or expanded form.

Expanded Form   **Standard Form**

_____ + _____ + _____ = 417

28. 600 + 50 + 4 = _____

COUNTING FORWARDS

29. Count forwards from 1 to 100. Write each missing number in the space provided.

    1   2 ___ ___ ___ ___ ___ ___ ___ ___

  11 ___ ___ ___ ___ ___ ___ ___ ___ ___

  21 ___ ___ ___ ___ ___ ___ ___ ___ ___

  31 ___ ___ ___ ___ ___ ___ ___ ___ ___

  41 ___ ___ ___ ___ ___ ___ ___ ___ ___

  51 ___ ___ ___ ___ ___ ___ ___ ___ ___

  61 ___ ___ ___ ___ ___ ___ ___ ___ ___

  71 ___ ___ ___ ___ ___ ___ ___ ___ ___

  81 ___ ___ ___ ___ ___ ___ ___ ___ ___

  91 ___ ___ ___ ___ ___ ___ ___ ___ ___

COUNTING BACKWARDS

30. Count backwards from 20 to 0. Write each missing number in the space provided.

    20 19 ___ ___ ___ ___ ___ ___ ___ ___

    10 ___ ___ ___ ___ ___ ___ ___ ___ ___

    0

SKIP COUNTING

31. Skip count by 2's to 100. Write each missing number in the space provided.

    2   4   ___ ___ ___ ___ ___ ___ ___ ___

    22 ___ ___ ___ ___ ___ ___ ___ ___ ___

    42 ___ ___ ___ ___ ___ ___ ___ ___ ___

    62 ___ ___ ___ ___ ___ ___ ___ ___ ___

    82 ___ ___ ___ ___ ___ ___ ___ ___ ___

32. Skip count by 5's to 100. Write each missing number in the space provided.

    5   10 ___ ___ ___ ___ ___ ___ ___ ___

    55 ___ ___ ___ ___ ___ ___ ___ ___ ___

33. Skip count by 10's to 100. Write each missing number in the space provided.

    10 20 ___ ___ ___ ___ ___ ___ ___ ___

34. Skip count by 25's to 100. Write each missing number in the space provided.

    25 ___ ___ ___

35. Skip count by 50's. Write the missing number in the space provided.

    50 ___

36. Skip count by 100's to 1,000. Write each missing number in the space provided.

    100 200 ___ ___ ___ ___ ___ ___ ___ ___

BEFORE, BETWEEN, AND AFTER NUMBERS (0-99)

37. In the spaces provided, write the missing before, between, and after numbers.

    _____ 8   9   31 _____ 33   57   58 _____

MONEY

Directions: Write the <u>name</u> (such as *nickel*) and <u>value</u> (such as 25¢) of each coin.
dime, half-dollar, nickel, penny, quarter

38.	39.	40.	41.	42.

Directions: Count the money and write the amount in the space provided. Use the symbols ¢ or $ in your answer.

43. _____

44. _____

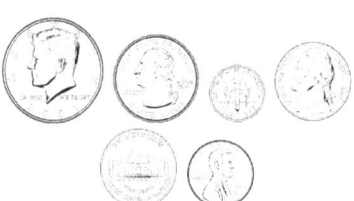

45. _____

46. Circle any combination of coins that are equivalent to 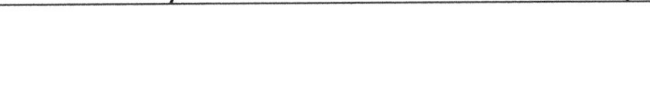 .

MONEY ADDITION AND SUBTRACTION (INCLUDING 2-DIGIT MONEY ADDITION AND SUBTRACTION WITH/WITHOUT REGROUPING)

Directions: Complete each money addition and subtraction sentence. Line up the numbers vertically before adding or subtracting.

47. 71¢ + 6¢ =	48. 54¢ + 26¢ =	49. 89¢ − 6¢ =	50. 62¢ − 4¢ =

Directions: Add the numbers (addends) with an addition sign. Subtract the numbers with a subtraction sign.

51. 62
    +25

52. 26
    + 8

53. 84
    −20

54. 71
    −43

TIME

Directions: What time is illustrated on the clock? When writing the time in digital form, remember to record the number for the little (hour) hand first; then write the number of minutes. If a second space is provided, use the words *o'clock*, *half past*, *quarter to*, or *quarter past* to write the time a different way.

55. _____

_____

56. _____

_____

57. _____
_____

58. _____
_____

59. _____

60. How many seconds are in one minute? _____

61. How many minutes are in one hour? _____

62. How many hours are in one day? _____

FRACTIONS AND PERCENT

63. In the fractions below, is ¼ or ¾ a larger fraction? How do you know?

_____

_____

Directions: What fraction of each circle is shaded? Write the fraction and use the initials **N** and **D** to mark the numerator and denominator.

64. 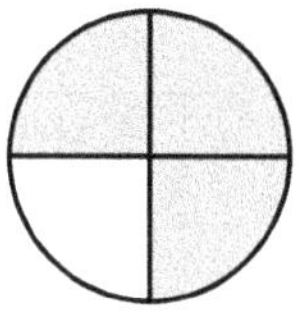	65. 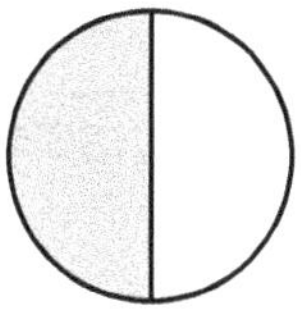	66. 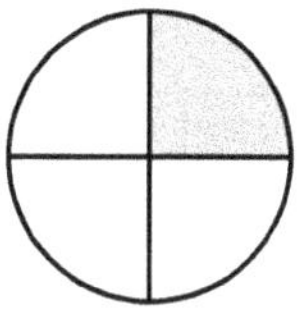	67. 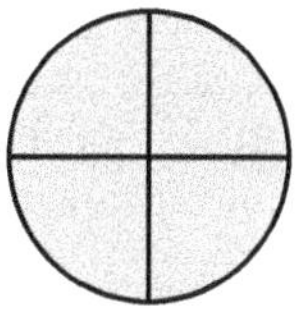
68. What % of this circle is shaded? _____	69. What % of this circle is shaded? _____	70. What % of this circle is shaded? _____	71. What % of this circle is shaded? _____

Name: _____   Date: _____

Elementary Math Test #2
(Grades: kindergarten, 1st, 2nd, 3rd, 4th, 5th)

CALENDAR WORD PROBLEMS (DAYS OF A WEEK, DAY OF THE MONTH, MONTHS OF A YEAR, YEARS)

1. Write the abbreviation for each day of the week in the space provided.

   Abbreviations: Fri., Mon., Sat., Sun., Thurs., Tues., Wed.

Sunday	Monday	Tuesday	Wednesday	Thursday	Friday	Saturday

Directions: Read each sentence and use a calendar to help solve the problem. Write the day of the week in the space provided.

2. What day of the week is between Monday and Wednesday? _____

3. What day of the week is before Friday? _____

4. What day of the week is after Friday? _____

5. If yesterday was Tuesday, what is today? _____

6. If yesterday was Tuesday, what will tomorrow be? _____

7. If today is Thursday, what was yesterday? _____

8. If today is Thursday, what will tomorrow be? _____

9. If today is Monday, what day of the week will it be 4 days from now? _____

10. If today is Friday, how many days are there until Sunday? _____

Directions: Read each sentence and use a calendar to help solve the problem. Write the ordinal number in the space provided. (Examples of ordinal numbers are 1st, 2nd, 3rd, 4th, and so on.)

11. What day of the month is between the 7th and 9th? _____

12. What day of the month is before the 29th? _____

13. What day of the month is after the 29th? _____

14. If yesterday was the 26th, what is today? _____

15. If yesterday was the 26th, what will tomorrow be? _____

16. If today is the 3rd, what was yesterday? _____

17. If today is the 3rd, what will tomorrow be? _____

18. If today is the 6th, what date will it be 8 days from now? _____

19. If today is the 14th, what date will it be 1 week from now? _____

20. If today is the 5th, how many days are there until the 12th? _____

21. Write the abbreviation for each month in the space provided.

   Abbreviations: Apr., Aug., Dec., Feb., Jan., Mar., Nov., Oct., Sept.

January	February	March	April	August	September	October	November	December

Directions: Read each sentence and use the listed months to help solve the problem. Write the month in the space provided.

January
February
March
April
May
June
July
August
September
October
November
December

22. What month is between October and December? _____

23. What month is before November? _____

24. What month is after November? _____

25. If the month is March, what will it be 2 months from now? _____

Directions: Read each sentence and write the correct year in the space provided.

26. What year is between 2022 and 2024? _____

27. What year was just before 2021? _____

28. What year is just after 2021? _____

29. If the year is 2021, what will it be 3 years from now? _____

30. If the year is 2021, what was it 5 years ago? _____

Directions: Use the information given below to determine if Ashley, Aaron, or Jenna is the oldest.

31. Ashley is older than Jenna, but Aaron is older than Ashley. _____

ADDITION WORD PROBLEMS

Directions: Read each sentence and draw a picture to help solve the problem. Write the addition fact in the spaces provided.

32. The farmer had 0 pigs. He bought 8. How many pigs does the farmer have now?

    _____ + _____ = _____

33. The farmer had 3 sheep. His ewe had 2 lambs. How many sheep does the farmer have now?

    _____ + _____ = _____

34. This morning, Alex milked 3 cows. Kailey milked 5. How many cows were milked altogether?

    _____ + _____ = _____

35. There were 8 chicks; 3 more hatched from their eggs. How many chicks in all?

    _____ + _____ = _____

36. Caleb caught 7 ducks. Jacob caught 3. How many ducks were caught altogether?

    _____ + _____ = _____

37. Ryan pulled up 10 carrots. Nathan pulled up 8. How many carrots were pulled up altogether?

    _____ + _____ = _____

38. Mandi peeled 5 potatoes. Taylor peeled 2. How many potatoes were peeled altogether?

    _____ + _____ = _____

39. Jessica picked 7 tomatoes from the vine. Joshua picked 9. How many tomatoes were picked altogether?

    _____ + _____ = _____

SUBTRACTION WORD PROBLEMS

Directions: Read each sentence and draw a picture to help solve the problem. Write the subtraction fact in the spaces provided.

40. The farmer has 9 pigs; 0 have been fed this morning. How many pigs need to be fed?

    _____ – _____ = _____

41. The farmer had 7 sheep. He sold 2 ewes. How many sheep does the farmer have left?

    _____ – _____ = _____

42. Sam has 5 cows to milk. Spencer milked 4 for him. How many cows still need to be milked?

    _____ – _____ = _____

43. Beth's class has been waiting for 8 chicks to hatch. This morning, they found that 4 hatched overnight. How many chicks are they still waiting for?

    _____ – _____ = _____

44. David and Michael saw 6 ducks. They caught 2. How many ducks got away?

    _____ – _____ = _____

45. I had 8 carrots. I gave Grace 1. How many carrots do I have left?

    _____ – _____ = _____

46. There were 9 potatoes growing in the ground. The farmer dug up 4 to take to market. How many potatoes are still in the ground?

    _____ – _____ = _____

47. There were 16 tomatoes growing on a vine. Eliana picked 5. How many tomatoes are left on the vine?

    _____ – _____ = _____

48. There were 19 ears of corn on the stalk. The farmer picked 10 to take to market. How many ears of corn are left on the stalk?

    _____ – _____ = _____

ADDITION AND SUBTRACTION WORD PROBLEMS

Directions: Read each sentence and draw a picture to help solve the problem. Write the addition or subtraction fact in the spaces provided.

49. There were 15 slices of ham. If 7 slices were eaten, how many slices are left?

    _____ = _____

50. Sophia has 1 red apple. Liam has 1 green apple. How many apples do they have altogether?

    _____ = _____

51. Noah caught 2 fish. His brother caught 7. How many fish did they catch altogether?

    _____ = _____

52. My aunt made 6 quilts. She sold 4. How many quilts does she have left?

    _____ = _____

53. Our football team has played 9 games so far. We lost 6 games. How many games did we win?

    _____ = _____

54. Grandma baked 4 pumpkin pies, and Mama baked 4 peach pies. How many pies were baked altogether?

    _____ = _____

55. Emma found 11 pecans. She fed 7 to the squirrels. How many pecans does Emma have left?

    _____ = _____

56. Lucas raked 6 bags of leaves. His dad raked up 1 more bag. How many bags of leaves in all?

    _____ = _____

57. I gathered 13 eggs from the barn. I accidentally broke 6. How many eggs do I have left?

    _____ = _____

MONEY ADDITION WORD PROBLEMS

58. Hailee had five nickels. She found 60¢ on her way to the store. How much money does she have now?

   _____

59. Judah has three dimes, two nickels, and six pennies. How much money does he have in all?

   _____

60. Elizabeth had four dimes. She earned 50¢ more. How much money does Elizabeth have now?

   _____

61. Emily had one quarter. Her mom gave her $3.00 more. How much money does she have now?

   _____

62. Bret wants to buy a set of baseball cards. It costs 96¢. He has 2 quarters and 8 nickels. Does he have enough money? (Show your work.)

   _____

63. Amishia wants to buy some candy. The cost is 84¢. She has 1 half-dollar, 3 dimes, and 1 penny. Does she have enough money? (Show your work.)

   _____

MONEY SUBTRACTION WORD PROBLEMS

64. Miranda had 48¢. She lost two nickels. How much money does she have left?

    _____

65. Jimmy has two dimes, four nickels, and five pennies. If he saves 45¢, how much can he spend?

    _____

66. Kim had 43¢. She spent two dimes at the store. How much money does she have left?

    _____

67. Sophie had $1.85. She gave her sister three quarters. How much money does she have left?

    _____

MONEY ADDITION AND SUBTRACTION WORD PROBLEMS

68. Mia had $1.25. She gave her brother one quarter. How much money does she have left?

   _____

69. Ava had two quarters. Her mom gave her $3.00 more. How much money does she have now?

   _____

70. Elijah had 88¢. He lost five nickels. How much money does he have left?

   _____

71. Logan has three dimes, four nickels, and one penny. If he saves 40¢, how much can he spend?

   _____

72. Olivia has four dimes, one nickel, and four pennies. How much money does she have in all?

   _____

73. Oliver had three dimes. He earned 10¢ more. How much money does Oliver have now?

   _____

74. Amelia had 84¢. She spent four dimes at the store. How much money does she have left?

   _____

Directions: Circle the correct change. Add the change and write the amount.

75. Jodi spent 68¢ at the store. She gave the store cashier $1.00. How much change did she receive?

   _____

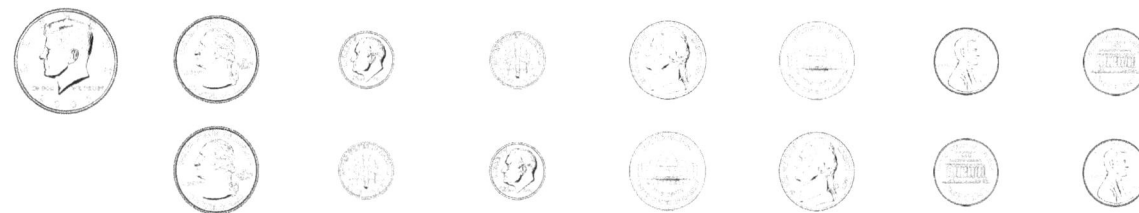

76. Derek spent $1.37 at the store. He gave the store clerk $5.00. How much change did he receive?

   _____

77. Kingston had $20.00. He spent $12.24 at the book fair. How much change did he receive?

   _____

TIME WORD PROBLEMS

Directions: Read each sentence and use a demonstration clock (and the time zone chart when necessary) to help solve the problem. Write the time in the space provided.

78. The farmer usually starts his chores at 6:30 A.M. Today he started them 30 minutes late. What time did he start his chores?

    _____

79. School starts at 8:30 A.M. Today, Shana was 15 minutes early. What time did she get to school?

    _____

80. Hunter usually eats lunch at 1:00 P.M. Today he ate it 45 minutes late. What time did he eat lunch?

    _____

81. The farmer usually goes to bed at 10:30 P.M. Today he went to bed 1 hour early. What time did he go to bed?

    _____

82. If it is 10:45 P.M., what time will it be 5 minutes later?

    _____

TIME ZONES					
Hawaii	Alaska	Pacific	Mountain	Central	Eastern
12:00	1:00	2:00	3:00	4:00	5:00

83. If it's 3:00 P.M. in Hawaii, what time is it in Colorado (Mountain time)?
   _____

84. If it's 7:30 A.M. in Alaska, what time is it in Hawaii? _____

85. If it's 1:00 P.M. in California (Pacific time), what time is it in Kansas (Central time)?
   _____

86. If it's 2:30 A.M. in New Mexico (Mountain time), what time is it in Washington (Pacific time)?
   _____

87. If it's 7:00 P.M. in Arkansas (Central time), what time is it in New York (Eastern time)?
   _____

88. If it's 12:30 A.M. in Florida (Eastern time), what time is it in Utah (Mountain time)?
   _____

89. If it's 5:00 P.M. in Alabama (Central time), what time is it in Tennessee (Central time)?
   _____

FRACTION WORD PROBLEMS

Directions: Read each sentence and draw a picture to help solve the problem. Write the fraction (and the % if asked for) in the space provided.

90. There were 4 pieces of apple pie. I ate 1 slice. What percentage of the pie did I eat?

    __ = _____

91. There were 2 pieces of peach pie. I ate 1 slice. What percentage of the pie did I eat?

    __ = _____

92. There were 4 pieces of pecan pie. We ate 3 slices. What percentage of the pie did we eat?

    __ = _____

93. There were 7 pieces of pumpkin pie. We ate 3 slices. How much of the pie did we eat?

    __

Name: _____ Date: _____

Elementary Math Test #3
(Grades: kindergarten, 1st, 2nd, 3rd, 4th, 5th)

Supplies: Ruler

PATTERNS (Shapes & Numbers)
1. Assign a capital letter to each shape. Begin with the letter *A* and assign the same letter to each matching shape. Then continue the pattern in the boxes.

   ★ ● ■ ■ ★ ● ■ ■ ☐ ☐ ☐ ☐

2. What would the next number of the pattern be?
   10, 15, 20, 25…
   _____

GREATER- AND LESS-THAN NUMBERS
3. Write **>**, **<**, or a greater- or less-than number in the space provided.

   2 ____ 5   _____ < 17   89 ____ 53

ORDERING NUMBERS FROM LEAST TO GREATEST (0-99)
4. In the spaces provided, write the numbers in order from least to greatest.
   24   6   19

   _____   _____   _____

ROUNDING NUMBERS TO THE NEAREST TEN AND HUNDRED
5. Round the numbers to the nearest ten.

   Number   **Rounded**

   65   _____

   17   _____

6. Round the numbers to the nearest hundred.
   Number   **Rounded**

   619   _____

   350   _____

7. According to the bar graph below, in what year did Elizabeth read the most number of books?

   _____

8. According to the line graph, in what year did Shana read the least number of books?

   _____

9. According to the bar graph, in what year did Elizabeth read 7 books?

   _____

10. According to the line graph, the number of books Shana read decreased between what two years?

    _____

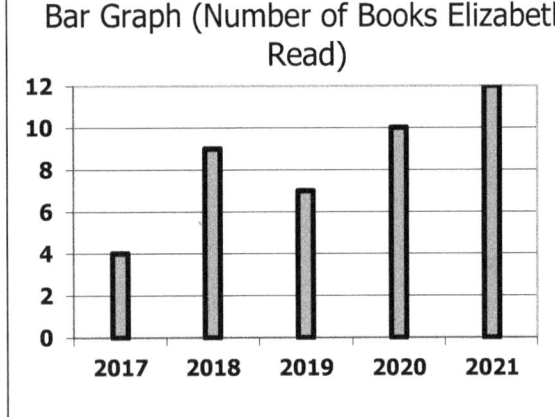

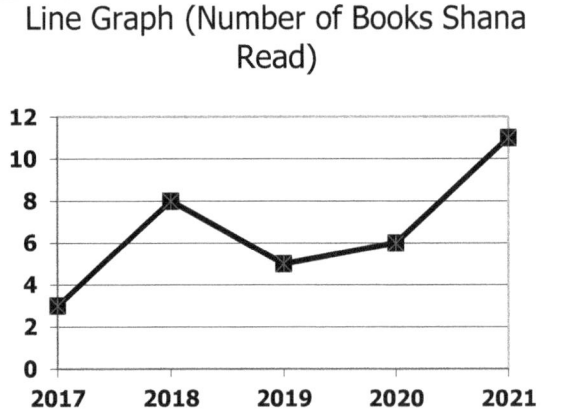

11. Use a ruler to make a line graph for the data below.

Year	Number of animals on Uncle Jim's farm
1995	8
2000	15
2005	21
2010	18
2015	25

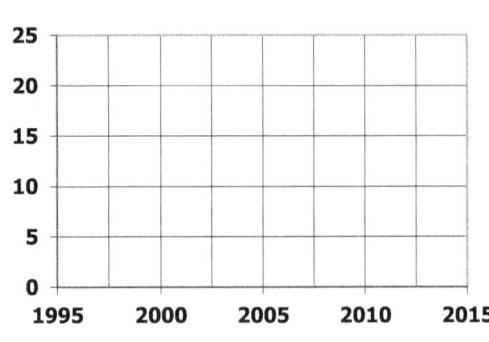

Find the perimeter.
12. Perimeter

    _____

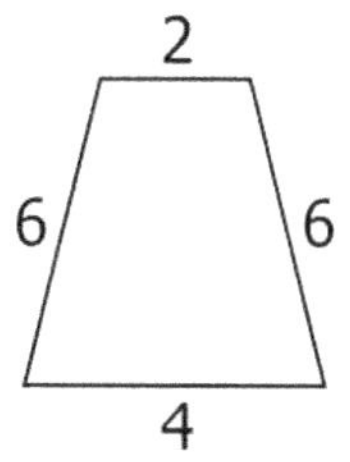

13. What is the area of the square to the right?

    _____ square units

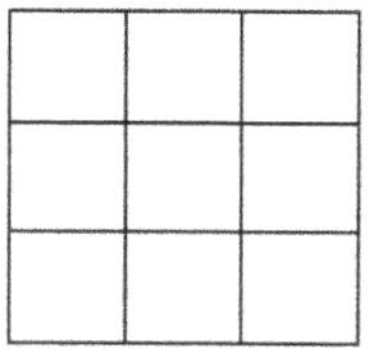

GEOMETRY
14. Circle the shape below that is **similar** to the shape in the box.

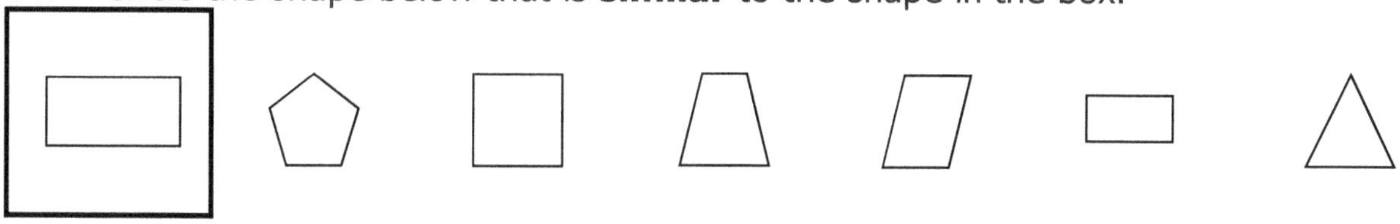

15. Circle the shape below that is **congruent** to the shape in the box.

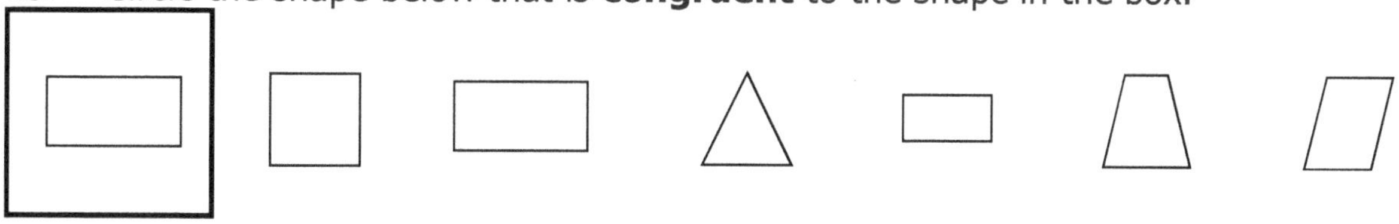

16. Use your ruler to show how many axes of symmetry (up to two) each figure has. Although some of the shapes have more than two lines of symmetry, illustrate the one or two most obvious ones. Note that one of the shapes doesn't have any lines of symmetry.

TEMPERATURE

Directions: What temperature is illustrated on the thermometer? Write the temperature in the space provided. Then circle the word that describes the temperature (use the guidelines below to help you).
Freezing: 32° or lower
Cold: 33°-63°
Warm: 64°-84°
Hot: 85° or higher

17. _____ freezing   cold   warm   hot

Directions: Read each sentence and use the thermometer to help solve the problem. (The thermometer does not illustrate the temperature from the word problems.) Write the temperature in the space provided.

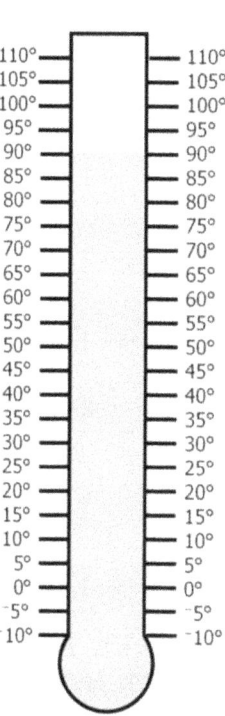

18. At 8:00 this morning, the temperature was 75°. By noon, it had risen 20°. What was the temperature at noon?

    _____

19. At 5:00 this afternoon, the temperature was 85°. By 9:00 P.M., it had dropped 5°. What was the temperature at 9:00 P.M.?

    _____

MEASUREMENT (LENGTH)

Directions: Write the measurement in the space provided (*1 foot*, *1 inch*, or *1 yard*).

20. 25 millimeters (mm) ≈ _____

21. 2½ centimeters (cm) ≈ _____

22. 12 inches (in.) = _____

23. 3 feet (ft.) = _____

24.	Use a ruler to measure the line in the box to the nearest inch. _____ inches _____
25.	Use a ruler to draw 25 millimeters in the box.
26.	Use a ruler to draw one centimeter in the box.
27.	Use a ruler to draw 1 inch in the box.

THIS MARKS THE END OF THE TEST FOR GRADES KINDERGARTEN THROUGH 2ND. STUDENTS IN GRADES 3RD THROUGH 5TH SHOULD CONTINUE.

CONVERSION PROBLEMS

Directions: Solve the conversion problems.

28.	6 feet = _____ inches
29.	24 in = _____ ft
30.	18 feet = _____ yards
31.	2 yd = _____ ft
32.	How many seconds are in six minutes? 6 minutes = _____ seconds
33.	How many minutes are in 180 seconds? 180 sec = _____ min
34.	How many hours are in 300 minutes? 300 minutes = _____ hours
35.	How many minutes are in six hours? 6 hr = _____ min

36. How many hours are in six days?

    6 days = _____ hours

37. How many days are in 48 hours?

    48 hours = _____ days

MULTIPLICATION AND DIVISION FACTS

SKIP COUNTING

38. Skip count by 2's to 18. Write each missing number in the space provided.

    __2__ _____ _____ _____ _____ _____ _____ _____ _____

39. Skip count by 3's to 27. Write each missing number in the space provided.

    __3__ _____ _____ _____ _____ _____ _____ _____ _____

40. Skip count by 4's to 36. Write each missing number in the space provided.

    __4__ _____ _____ _____ _____ _____ _____ _____ _____

41. Skip count by 5's to 45. Write each missing number in the space provided.

    __5__ _____ _____ _____ _____ _____ _____ _____ _____

42. Skip count by 6's to 54. Write each missing number in the space provided.

    __6__ _____ _____ _____ _____ _____ _____ _____ _____

43. Skip count by 7's to 63. Write each missing number in the space provided.

    __7__ _____ _____ _____ _____ _____ _____ _____ _____

44. Skip count by 8's to 72. Write each missing number in the space provided.

    __8__ _____ _____ _____ _____ _____ _____ _____ _____

45. Skip count by 9's to 81. Write each missing number in the space provided.

   9   ____   ____   ____   ____   ____   ____   ____   ____

Directions: Solve.

46. 8 × 2 = ____   49. 18 ÷ 6 = ____

47. 5 × 1 = ____   50. 4 ÷ 1 = ____

48. 7 × 0 = ____

Directions: What does 2 times 3 equal (2 × 3 = __). Use stars or other shapes and the 2 cells below to illustrate the answer.

51. 2 × 3 = ____

Directions: What does 8 divided by 2 equal (8 ÷ 2 = __). Divide the 8 stars between the 2 cells below to illustrate the answer. You can cross out the stars as you use them.

52. 8 ÷ 2 = ____

MULTIPLICATION WORD PROBLEMS

Directions: Read each sentence and draw a picture to help solve the problem. Write the multiplication fact in the spaces provided.

53. Each of the 15 farmers have 0 white pigs. How many white pigs do they have in all?

    _____ × _____ = _____

54. Each of the farmer's 4 ewes had 2 lambs. How many lambs in all?

    _____ × _____ = _____

55. This morning, 5 children milked 1 cow apiece. How many cows were milked altogether?

    _____ × _____ = _____

56. Each of the 4 mother hens have 7 eggs that have started to hatch. How many chicks will there be in all?

    _____ × _____ = _____

57. Three children caught 5 ducks apiece. How many ducks were caught altogether?

    _____ × _____ = _____

58. Seven children pulled up 7 carrots apiece. How many carrots were pulled up altogether?

    _____ × _____ = _____

59. Nine children peeled 6 potatoes apiece. How many potatoes were peeled altogether?

    _____ × _____ = _____

60. Two children picked 5 tomatoes from the vine. How many tomatoes were picked altogether?

    _____ × _____ = _____

DIVISION WORD PROBLEMS

Directions: Read each sentence and draw a picture to help solve the problem. Write the division fact in the spaces provided.

61. There were 28 carrots; 7 children divided them. How many carrots did each child get?

    _____ ÷ _____ = _____

62. There were 12 celery sticks. Cara and Pam divided them. How many sticks did each one get?

    _____ ÷ _____ = _____

63. Six potatoes need to be peeled; 2 children divided them. How many potatoes did each child peel?

    _____ ÷ _____ = _____

64. Eighteen tomatoes were picked from the vine; 3 children divided them. How many tomatoes did each child get?

    _____ ÷ _____ = _____

MULTIPLICATION AND DIVISION WORD PROBLEMS

Directions: Read each sentence and draw a picture to help solve the problem. Write the multiplication or division fact in the spaces provided.

65. There were 21 slices of ham. If 7 people divided them, how many slices did each person get?

    _____ = _____

66. There were 24 red apples. Eight children divided them. How many apples did each child get?

    _____ = _____

67. Five women were selling quilts at the flea market. If each woman sold 6 quilts, how many quilts were sold altogether?

    _____ = _____

68. Nine children were fishing. If each child caught 4 fish, how many fish were caught altogether?

    _____ = _____

69. There were 5 slices of pumpkin pie. Five people divided them. How many slices did each person get?

    _____ = _____

70. Two children gathered 2 eggs apiece from the barn. How many eggs were gathered altogether?

    _____ = _____

71. Four brothers found 32 turkey feathers in their yard. If they divide them evenly, how many feathers will each brother get?

    _____ = _____

72. Seven children were raking leaves. If each child raked up 5 bags full, how many bags of leaves were there?

    _____ = _____

73. Abigail had 30 pecans. She divided them between 5 squirrels. How many pecans did each squirrel get?

    _____ = _____

Name: _____   Date: _____

Elementary Math Test #4
(Grades: 3rd, 4th, 5th)

Directions: Show all your work in the boxes provided.

1. In an ordered pair such as (2, 4), 2 represents the _____-coordinate, and 4 represents the _____-coordinate. (Fill in the blanks with x or y.)

Directions: Graph the following ordered pairs in the coordinate plane below.

2. $A(7, -7)$

3. $B(-1, 0)$

4. $C(0, 7)$

5. $D(-5.5, 2½)$

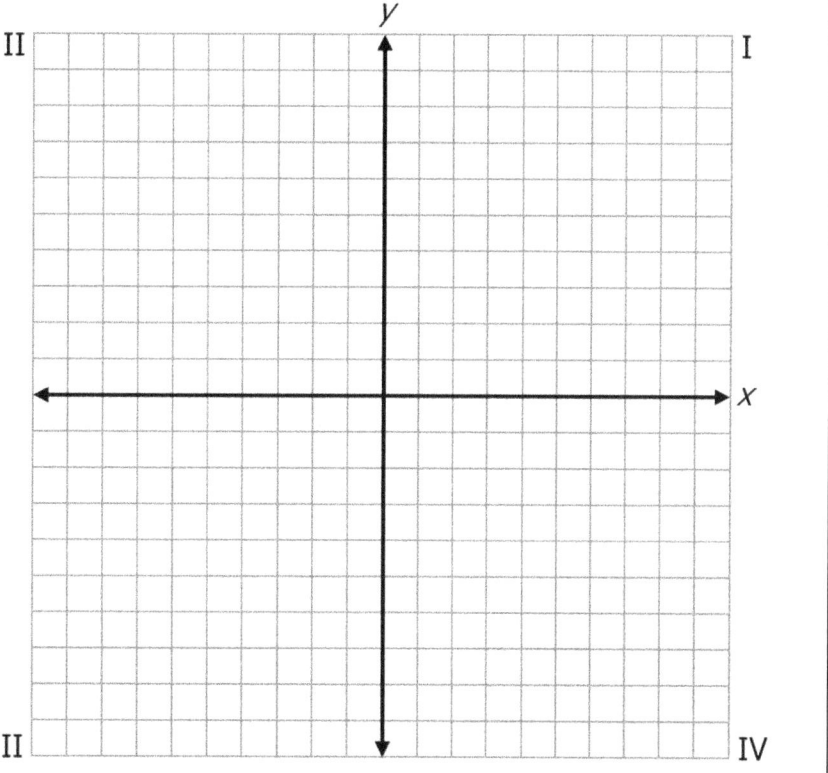

6. Circle the origin in the coordinate plane above.

7. The four sections that a coordinate plane is divided into by the x- and the y-axes are called _____. These are numbered I, II, III, and IV.

43
Copyright © 2019-2022 by 70 Times 7 Math (a division of Habakkuk Educational Materials). All rights reserved.

8. Sketch a standard (x, y) coordinate plane in the box to the right and use it to find the location of a square's corner. (The three known corners of the square are listed below.) Record the point of the fourth corner in the space provided.
(−3, −3), (3, 3), (3, −3)

_____

Addition and Subtraction (with regrouping)

Directions: Rewrite the problems vertically to solve.

9. 236 + 45 = _____	10. 6943 + 728 = _____
11. 72 − 54 = _____	12. 403 − 35 = _____

Multiplication (with and without regrouping)

Directions: Rewrite the problems vertically to solve.

13. 63 × 4 = _____	14. 32 × 46 = _____

15. 421 × 213 = _____

16. 43 × 20 = _____

17. 241 × 302 = _____

True or False
18. You can change a remainder to a fraction by writing the remainder over the divisor.

   _____

19. 37 ÷ 5 = _____

20. Rewrite the remainder in 37 ÷ 5 as a fraction.

Long Division

21. 72 ÷ 2 = _____

22. 890 ÷ 5 = _____

23. 195 ÷ 15 = _____

Division (0 Remaining)

Directions: Turn the remainder into a decimal.

24. 7 ÷ 4 = _____

Directions: Divide the decimals.

Division (Decimal Dividend)

25. 47.1 ÷ 3 = _____

Decimal Dividend and Divisor

26. 6.32 ÷ 0.4 = _____

Decimal Dividend and Divisor

27. 0.186 ÷ 0.02 = _____

Decimals in Addition and Subtraction Problems

Directions: Line up the decimals to solve.

28. 7.2 + 61.3 = _____	29. 18.69 − 5.24 = _____
30. 9.5 − .35 = _____	31. 679.1 − 85 = _____

Decimals in Multiplication

Directions: Rewrite the problems vertically to solve.

32. 7.4 × 0.2 = _____	33. 0.15 × 0.3 = _____
34. 1.742 × 0.5 = _____	

35. On the top half of the chart below, fill in the missing word name.

36. On the bottom half of the chart below, fill in the missing number.

thousands	hundreds	tens	ones		hundredths	thousandths	ten thousandths
1,000	100	10	1		$\frac{1}{100}$	$\frac{1}{1,000}$	$\frac{1}{10,000}$

Directions: Rewrite each expanded numeral as a decimal.

37. $5(1,000) + 8(100) + 4(10) + 9(1) + 6(\frac{1}{10}) + 3(\frac{1}{100}) + 2(\frac{1}{1,000})$

   _____

38. $8(1,000) + 3(100) + 9(10) + 5(\frac{1}{10}) + 7(\frac{1}{100}) + 4(\frac{1}{1,000}) + 6(\frac{1}{10,000})$

   _____

39. $5(\frac{1}{10,000}) + 4(100) + 7(1,000) + 2(\frac{1}{100}) + 6(\frac{1}{10}) + 3(10) + 1(\frac{1}{1,000})$

   _____

Directions: Rewrite each word number as a decimal.

40. Three hundred forty-two thousandths   _____

41. Six hundred twenty-seven ten thousandths   _____

42. Four thousand seven hundred nineteen and six hundred fifty-two thousandths

   _____

Directions: Use the symbol <, >, or = to compare the decimals below.

43. 7.2043 _____ 7.206

44. 5.2 _____ 5.20

Directions: What fraction of each circle is shaded? Write the fraction in the space provided. (If the fraction can be reduced, write it after the equal sign.) Then use the initials **N** and **D** to mark the numerator and denominator.

45. _____ = _____	46. _____ = _____	47. _____ = _____
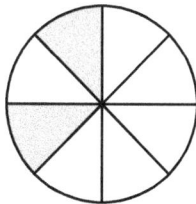	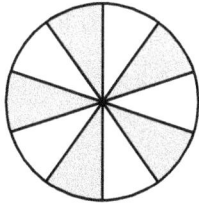	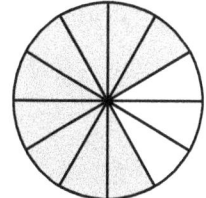
48. What percent of this circle is shaded? _____	49. What percent of this circle is shaded? _____	50. What percent of this circle is shaded? _____

51. In the fraction ¾, 3 is the _____, and 4 is the _____.

Proper and Improper Fractions

52. Give an example of a proper fraction.

53. Give an example of an improper fraction.

Directions: Rewrite the fractions in lowest terms or as integers.

54. $\dfrac{0}{8} =$

55. $\dfrac{9}{9} =$

56. $\dfrac{5}{-5} =$

57. $\dfrac{3}{1} =$

58. $\dfrac{32}{4} =$

59. $\dfrac{6}{21} =$

Directions: Rewrite the fraction as a mixed number. Reduce to lowest terms.

60. $\dfrac{36}{10} =$

Directions: Rewrite the mixed number as an improper fraction.

61. $5\dfrac{3}{4} =$

Adding and Subtracting Fractions (Like Denominators)

Directions: Identify the sum or difference. (Always remember to reduce fractions to lowest terms and to rewrite improper fractions as mixed numbers.)

62. $\dfrac{5}{9} + \dfrac{4}{9} =$

63. $\dfrac{6}{8} - \dfrac{2}{8} =$

64. $\dfrac{12}{6} - \dfrac{4}{6} =$

65. $\dfrac{6}{17} + \dfrac{1}{17} + \dfrac{4}{17} =$

Adding and Subtracting Fractions (Different Denominators)

Directions: When adding or subtracting fractions with different denominators, rewrite the problems vertically to solve. After identifying the sum or difference, remember to reduce fractions to lowest terms and to rewrite improper fractions as mixed numbers.

66. $\dfrac{4}{12} + \dfrac{1}{6} =$

67. $\dfrac{3}{6} - \dfrac{2}{9} =$

68. $\begin{array}{r} \dfrac{2}{10} \\ -\dfrac{3}{5} \\ \hline \end{array}$

69. $\dfrac{3}{4} + \dfrac{5}{8} + \dfrac{1}{2} =$

Adding and Subtracting Mixed Numbers (Like Denominators)

Directions: Add or subtract the mixed numbers. Reduce to lowest terms.

70. $2\frac{4}{7} + 3\frac{2}{7} =$

71. $4\frac{5}{8} - 1\frac{3}{8} =$

72. $5\frac{6}{8} + 2\frac{3}{8} =$

Adding and Subtracting Mixed Numbers (Different Denominators)

Directions: Add or subtract the mixed numbers.

73. $2\frac{5}{6} + 6\frac{9}{12} =$

74. $\begin{array}{r} 8 \\ -5\frac{4}{9} \\ \hline \end{array}$

75. $\begin{array}{r} 6\frac{1}{4} \\ -5\frac{2}{6} \\ \hline \end{array}$

Directions: Multiply the fractions and reduce to lowest terms.

76. $\frac{4}{5} \times \frac{3}{6} =$

77. $\left(\frac{1}{6}\right)\left(\frac{9}{3}\right) =$

78. $-\dfrac{7}{9} \times \dfrac{4}{5} =$

79. $1\dfrac{3}{6} \times 2 =$

Directions: Cancel out the fractions before multiplying.

80. $\dfrac{3}{9} \times \dfrac{5}{14} \times \dfrac{7}{4} =$

Directions: Record the fraction form and the reciprocal in the box below.

Number	In Fraction Form	The Reciprocal
81. $\dfrac{2}{5}$		
82. 4		
83. $2\dfrac{1}{5}$		

84. If you wanted to find the reciprocal of ¾, what would you multiply ¾ by? _____

What would the product be? _____

Directions: Divide the fractions. Reduce to lowest terms.

85. $\dfrac{3}{10} \div \dfrac{3}{5} =$

86. $2 \div 3\dfrac{1}{2} =$

Directions: Graph the fractions. Remember to insert the markers an equal distance apart.

87. Graph $\frac{1}{3}$.

88. Graph $\frac{-5}{3}$.

89. Graph $\frac{5}{3}$.

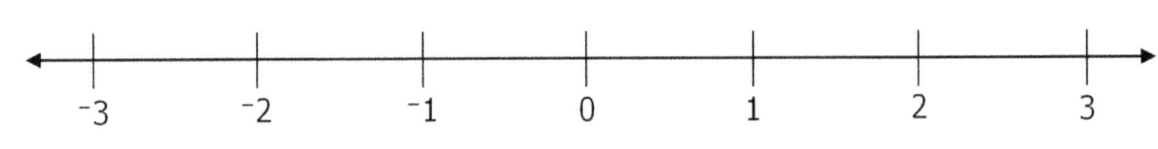

Name: _____   Date: _____

Elementary Math Test #5
(Grades: 3rd, 4th, 5th)

Supplies: Calculator

Directions: Show all your work in the boxes provided.

Reading Numbers
Directions: Read each number and record it in the space provided.

1. 4 hundred & 21   _____

2. 57 million, 9 hundred & 31 thousand, 4 hundred & 22   _____

3. 6 thousand, 3 hundred & 48   _____

4. seven   _____

5. 2 billion, 3 hundred & 15 million, 5 hundred & 76 thousand, 1 hundred & 49
   _____

6. 74 thousand, 9 hundred & 81   _____

7. 2 hundred & 54 thousand, 7 hundred & 89   _____

8. 4 hundred & 51 million, 3 hundred & 75 thousand, 2 hundred & 60
   _____

9. forty-three   _____

10. 7 million, 5 hundred & 83 thousand, 4 hundred & 7   _____

Directions: Multiply 0.41 by $9 million.

11. (0.41)($9 million) = _____ or _____

Rounding Numbers and Decimals

Round 394,572 to the nearest
12. 10   _____
13. 100   _____
14. 1,000   _____
15. 10,000   _____
16. 100,000   _____

Round 3.8556 to the nearest
17. whole number   _____
18. tenth   _____
19. hundredth   _____
20. thousandth   _____

21. Round 672.597 to the nearest hundredth.   _____

Directions: Round the mixed number to the nearest whole number.

22. $5\frac{3}{4}$ = _____

Directions: Divide the numbers and round the decimal quotient to the nearest hundredth. <u>Do not use a calculator.</u>

23. 11 ÷ 8 = _____

Directions: <u>Estimate</u> by rounding the decimals to the nearest whole number. Choose a slightly larger or smaller number for the dividend (the first number) if necessary so that there is no remainder in your estimation.

24. 11.7 ÷ 4.3 ≈ _____

25. 21.34 ÷ 9.7 ≈ _____

The Metric System

Directions: List the below prefixes in order from the greatest value to the least.
    centi- (cm), deci- (dm), deka- (dkm), hecto- (hm), kilo- (km), milli- (mm)

26.	27.	28.	**METER (m)**	29.	30.	31.

Directions: Convert the units of measurement.

32. 5.4 kilometers = _____ meters

33. 723 millimeters = _____ meters

Directions: Choose a number or number name from the boxes below for 34-39.

1,000	100	10	one tenth (0.1)	one hundredth (0.01)	one thousandth (0.001)

34. hectometer = _____ meters

35. decameter = _____ meters

36. centimeter = _____ of a meter

37. millimeter = _____ of a meter

38. kilometer = _____ meters

39. decimeter = _____ of a meter

Approximate Conversions
Directions: Solve the conversion problems.

40. 4 inches ≈ _____ millimeters

41. 4 inches ≈ _____ centimeters

42. 4 inches ≈ _____ meters

43. 5 feet ≈ _____ centimeters

44. 5 feet ≈ _____ meters

45. 6 yards ≈ _____ centimeters

46. 6 yards ≈ _____ meters

Measurements (Liquids)

47. 1 tablespoon = _____ teaspoons (tsp.)

48. 1 ounce = _____ tablespoons (Tbsp.)

49. 1 cup = _____ ounces (oz.)

50. 1 pint = _____ cups

51. 1 quart = _____ cups   or   _____ pints (pt)

52. 1 gallon = _____ quarts (qt)

Conversion Problems

Directions: Solve the conversion problems.

53. 2 tablespoons = _____ teaspoons

54. 9 tsp = _____ Tbsp

55. 4 tablespoons = _____ ounces

56. 3 oz = _____ Tbsp

57. 2 cups = _____ ounces

58. 24 ounces = _____ cups

59. 4 cups = _____ pints

60. 3 pints = _____ cups

61. 2 quarts = _____ pints

62. 6 pints = _____ quarts

63. 12 quarts = _____ gallons

64. 4 gal = _____ qt

Measurements (Weight)
65. Describe the weight of a gram (g) and tell how many milligrams are in one gram.
_____
_____

66. 16 ounces = _____

67. kilogram (kg) = _____

Conversion Problems

Directions: Solve the conversion problems.

68. 2 pounds = _____ ounces

69. 48 ounces = _____ pounds

70. 2,000 grams = _____ kilograms

71. 3 kilograms = _____ grams

Directions: Fill in the blanks.
72. According to the Fahrenheit temperature scale, water freezes at _____.

73. According to the Fahrenheit temperature scale, water boils at _____.

74. According to the Fahrenheit temperature scale, normal is _____.

75. The formula used to convert Fahrenheit into Celsius is _____.

76. Use your calculator to convert 212° Fahrenheit into Celsius.
    _____ C

Directions: Fill in the blanks.

77. According to the Celsius temperature scale, water freezes at _____.

78. According to the Celsius temperature scale, water boils at _____.

79. According to the Celsius temperature scale, normal is _____.

80. The formula used to convert Celsius into Fahrenheit is _____.

81. Convert 100° Celsius into Fahrenheit.
    _____ F

Least Common Multiple (LCM)

Directions: Identify the first nine nonzero multiples of the numbers 3 and 7. Then find the least common multiple (LCM) of 3 and 7.

82. Multiples of 3: _____

83. Multiples of 7: _____

84. What is the LCM of 3 and 7? _____

Greatest Common Factor (GCF)

85. If you are looking for the greatest common factor (GCF) of two numbers, such as 6 and 12, you are looking for what?
    _____

86. What are the factors of 10? _____

87. What are the factors of 20? _____

88. What are the common factors of 10 and 20? _____

89. What is the GCF of 10 and 20? _____

90. Are the numbers 10 and 20 **relatively prime**? Tell how you know.
    _____

Can the following numbers be divided evenly by 2, by 5, by 10, or by all three of the numbers?

91. The number 75 is divisible by _____.

92. The number 54 is divisible by _____.

93. The number 500 is divisible by _____.

Prime and Composite Numbers

94. Is the number **10** prime, composite, or neither prime nor composite?
    _____

95. Is the number **13** prime, composite, or neither prime nor composite?
    _____

96. Is the number **1** prime, composite, or neither prime nor composite?
    _____

True or False
97. A number less than 2 (such as −2, −1, 0, or 1) cannot be prime or composite.
    _____

98. Use factor trees to find the prime factorization of number 18. Use exponents whenever possible.

   _____

99. What are the prime factors of 42? _____

Directions: Use factor trees to locate the GCF and the LCM of each pair of numbers below.

100. GCF of 10 and 15: _____

101. LCM of 10 and 15: _____

102. GCF of 42 and 70: _____

103. LCM of 42 and 70: _____

Name: _____ Date: _____

Elementary Geometry Test #6
(Grades: 3rd, 4th, 5th)

Supplies: Protractor

Directions: Show all your work in the boxes provided.

Geometry studies points, lines, and planes, and the shapes they create. Sketch these figures in the spaces provided.

1. Point	2. Line	3. Plane

Directions: Identify the lines below. Be as specific as possible.
intersecting lines, parallel lines, perpendicular lines

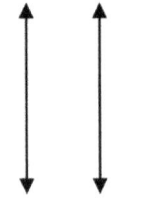   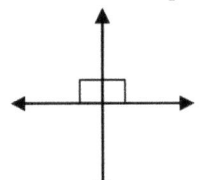   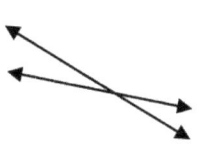   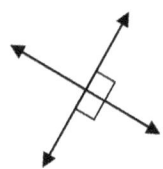

4. _____   5. _____   6. _____   7. _____

8. What do the arrowheads of a line signify?
   _____

Directions: Fill in the blanks.

9. A _____ has at least two points.

10. A _____ is flat. It has at least three points that are not all on the same line.

11. _____ has at least four points that are not all lying in the same plane.

12. _____ lines do not intersect, and they are the same distance apart.

Directions: Classify each figure in the box as a *simple curve*, a *simple closed curve*, a *closed curve*, or a *curve*.

13.	14.	15.	16.
∪	△	↩	8

17. How do you distinguish curves from simple curves? Use the illustrations above to help you.

   _____

   _____

Directions: Identify the two polygons below as simple or complex. Tell how you know the polygon is simple or complex.

18. _____   19. _____

   _____   _____

20. Cross out the shapes that are not polygons.

21. The first shape above is a polygonal curve, but is it also a polygon? Explain your answer.

   _____

22. Which shape above is an irregular polygon (a polygon that is not <u>regular</u>)? Explain your answer.

   _____

   _____

23. What does *equilateral* mean?

   _____

24. What does *equiangular* mean?

   _____

25. What are the two dimensions of a plane (flat) shape?

   _____

26. What are the three dimensions of a 3-D (solid) figure?

   _____

Directions: Record the number of sides, angles, and vertices each polygon has.
   3, 4, 5, 6, 7, 8, 9, 10, 11, 12, or ***unknown***

27. nonagon   _____   33. heptagon   _____

28. dodecagon   _____   34. quadrilateral   _____

29. octagon   _____   35. decagon   _____

30. pentagon   _____   36. triangle   _____

31. hendecagon   _____   37. hexagon   _____

32. *n*-gon   _____   38. What is a 5-gon? _____

39. A trapezoid has one pair of parallel opposite sides, called *bases*. Put an *x* through the parallel sides in the trapezoid below.

40. Fill in the blanks: A _____ is a parallelogram with all four sides the same length. A _____ is a rectangle with all four sides the same length.

41. Circle the concave shapes below.

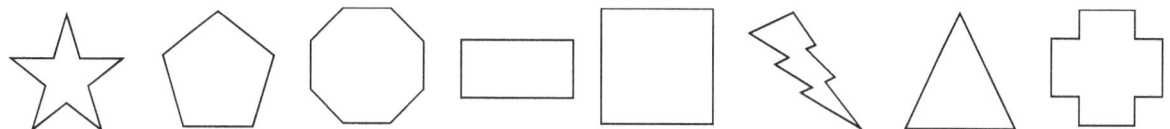

42. How do you know that the shapes you circled are concave rather than convex?
   _____
   _____

Directions: Identify the number of faces each polyhedron has.
   4, 5, 6, 7, 8, 10

43. Octahedron _____

44. Hexahedron _____

45. Pentahedron _____

46. Decahedron _____

47. Tetrahedron _____

48. Heptahedron _____

Directions: Write the name of each flat shape in the space provided. Use a word from the list.

circle, decagon, dodecagon, hendecagon, heptagon, hexagon, nonagon, octagon, oval, parallelogram, pentagon, rectangle, rhombus, square, trapezoid, triangle

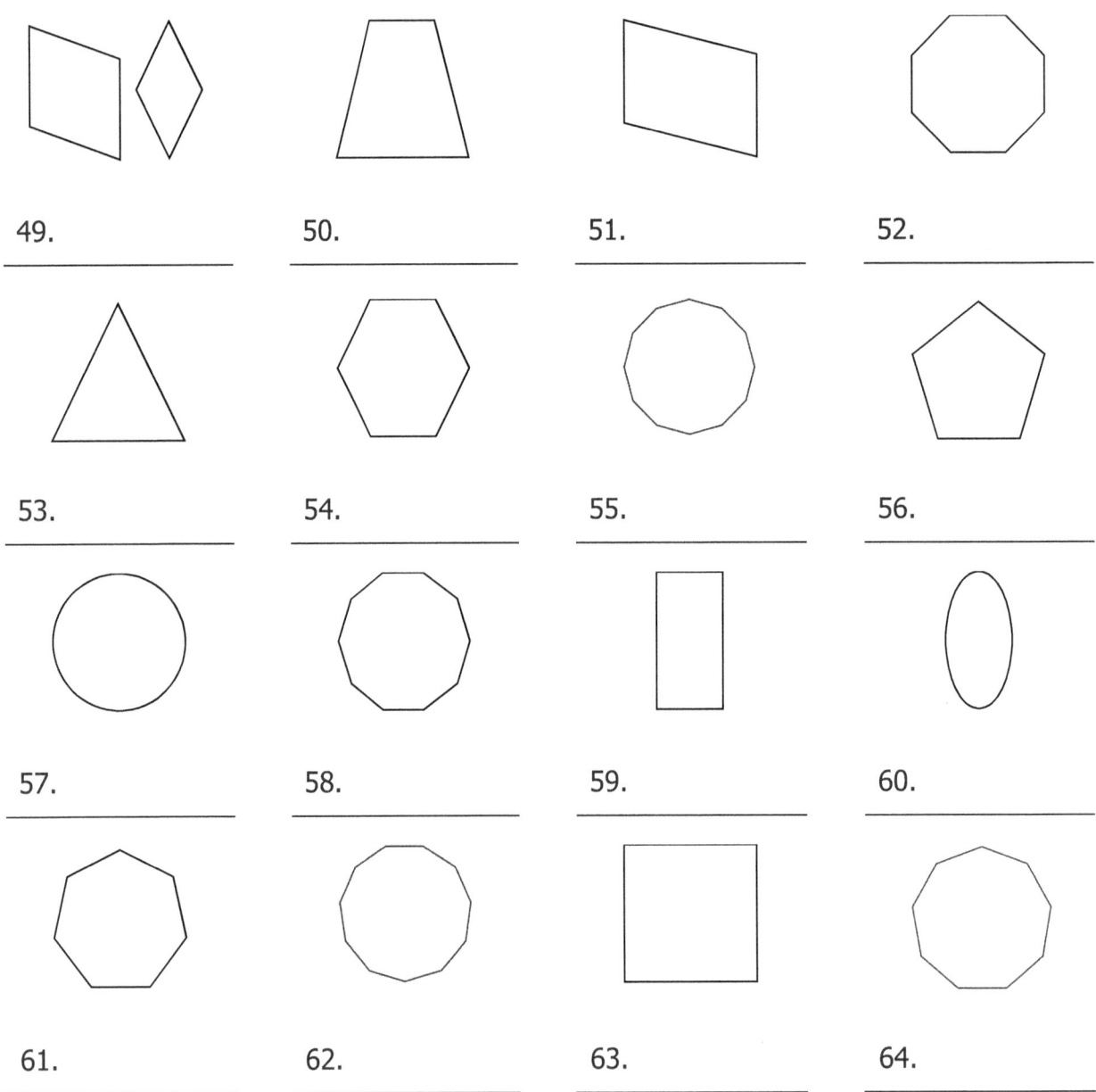

49. _____ 50. _____ 51. _____ 52. _____

53. _____ 54. _____ 55. _____ 56. _____

57. _____ 58. _____ 59. _____ 60. _____

61. _____ 62. _____ 63. _____ 64. _____

65. Circle all the shapes on this page that are also quadrilaterals.

Directions: Write the name of each solid figure in the space provided. Use a word from the list.

cone, cube, cylinder, hexagonal prism, hexagonal pyramid, pentagonal prism, pentagonal pyramid, rectangular prism (or cuboid), sphere, triangular prism, triangular pyramid

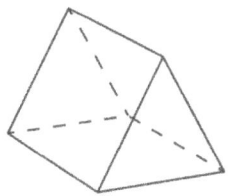 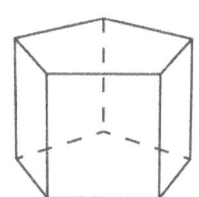 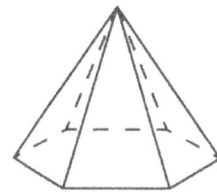

66. _____   67. _____   68. _____

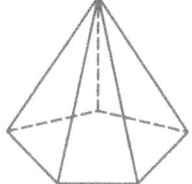 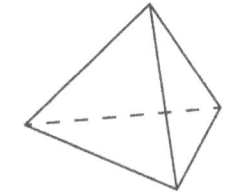 

69. _____   70. _____   71. _____

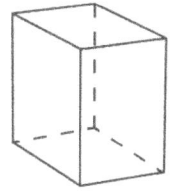 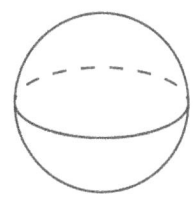 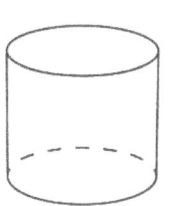

72. _____   73. _____   74. _____

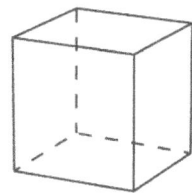 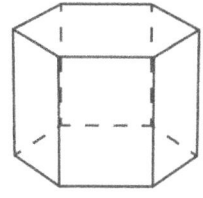

75. _____   76. _____

77. Which two shapes above are also called heptahedrons?

_____

Directions: Fill in the blanks. Use each of the choices once and be as specific as possible.

pyramid, geometric solid (or *three-dimensional figure*), *polyhedron, prism*

78. A _____ is not a flat shape.

79. A 3-D figure that has polygons for faces is a _____.

80. A _____ has two congruent polygonal faces, called bases, and all the rest of its faces are quadrilaterals.

81. A _____ has only one polygonal base, and the rest of its faces are triangles.

82. Why is a cylinder not a prism or a polyhedron? _____
    _____

83. Euler's Formula, which is true for any convex polyhedron, is _____.

Directions: Identify the number of vertices, edges, and faces the figure has. Then compare Euler's Formula with the number of vertices, edges, and faces.

84. Vertices: _____
85. Edges: _____
86. Faces: _____
87. Euler's Formula:
    _____

Directions: Identify the angle. Write *right, acute, obtuse, straight.,* or *zero angle*.

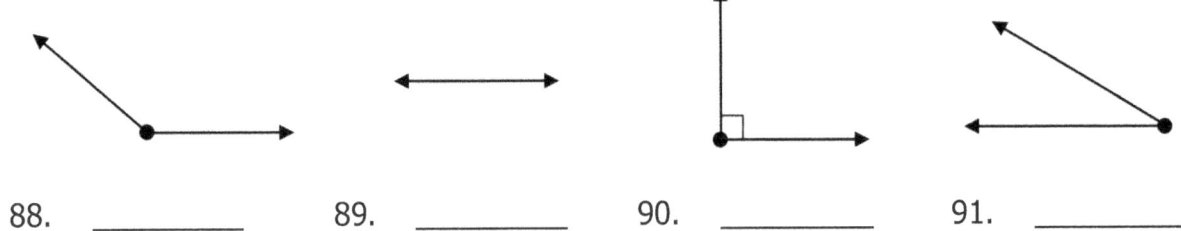

88. _____ 89. _____ 90. _____ 91. _____

Directions: Complete the sentences.

92. An **acute angle** measures _____.

93. A **right angle** measures _____.

94. An **obtuse angle** measures _____.

95. A **straight angle** measures _____.

96. I estimate the measure of this angle to be _____.
    A. 35°   C. 95°
    B. 85°   D. 115°

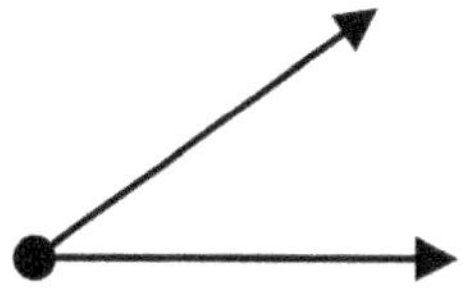

On this test, your protractor may only be used for drawing the angle.

97. In the box below, use your protractor to draw a 70° angle. Name the angle you drew ∠PQR.

98. In the space provided, list two other ways of naming ∠PQR.

    _____ and _____

99. What notation could you use to show that the angle measures 70°?

    _____

100. What is the vertex of the 70° angle you drew? _____

101. What are the rays of the 70° angle you drew? Use the correct notation.

     _____ and _____

Directions: Find the <u>area</u> of the square, the rectangle, and the parallelogram.

Square	Rectangle	Parallelogram
102. _____ in²	103. _____ in²	104. _____ ft²
5 in. (side)	3 in. × 7 in.	5 ft., 3 ft. height

105. Fill in the blanks: A radius is _____ the length of the diameter.

   (A radius goes _____ through a circle.)

106. If the diameter of a circle is 14 inches, what is its radius?

   _____

   14 in.

107. This circle has a radius of 2 cm. What is its diameter?

   _____

   2 cm

108. Do reflections (flips), rotations (turns), and translations (slides) change in position, size, or shape?

   _____

Directions: Identify the type of transformation below—*reflection (flip)*, *rotation (turn)*, or *translation (slide)*.

109. _____	110. _____	111. _____

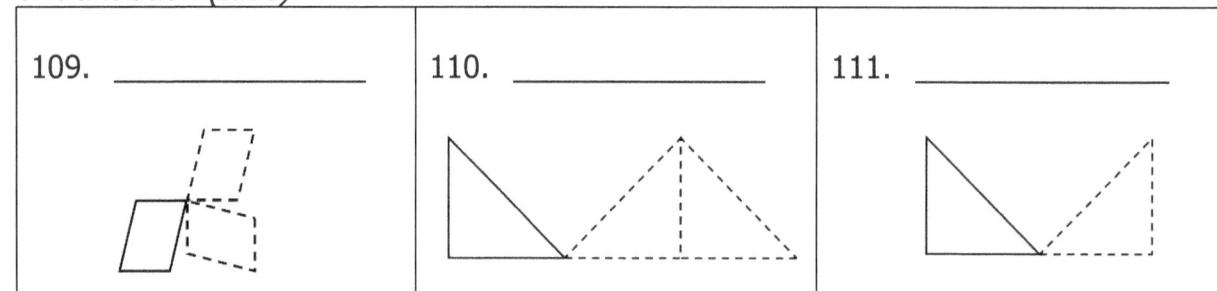

Name: _____   Date: _____

Key for Beginning Geometry Classwork and Test
(Grades: kindergarten, 1st, 2nd)
(Circle one: Classwork #1, #2, #3, #4, Test #1)

Text: *Polygons, Polyhedrons, and Other Shapes for Grades Pre-K Through 5th*

Record the number of sides, angles, and vertices each polygon has:
   3, 4, 5, 6, 7, 8, 9, 10, or *unknown*.

1. nonagon   **9**   6. quadrilateral   **4**
2. octagon   **8**   7. decagon   **10**
3. pentagon   **5**   8. triangle   **3**
4. heptagon   **7**   9. hexagon   **6**
5. *n*-gon   **unknown**   10. What is a 3-gon? **triangle**

Identify the number of faces each polyhedron has:
   4, 5, 6, 7, 8, or 10.

11. octahedron   **8**   14. decahedron   **10**
12. hexahedron   **6**   15. tetrahedron   **4**
13. pentahedron   **5**   16. heptahedron   **7**

17. Cross out the shapes that are not polygons.

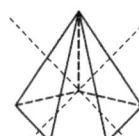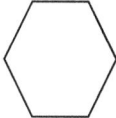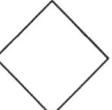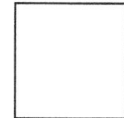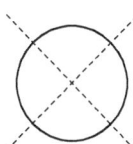

18. Circle all the quadrilaterals on the next page. **The parallelogram, trapezoid, diamond/rhombus, rectangle, and square should be circled.**

19. Which shape on the next page can be called a decagon? **star**

73

Directions: Write the name of each flat shape in the space provided. The first one has been done for you.

circle, cross, diamond/rhombus, heart, heptagon, hexagon, nonagon, octagon, oval, ~~parallelogram~~, pentagon, rectangle, square, star, trapezoid, triangle

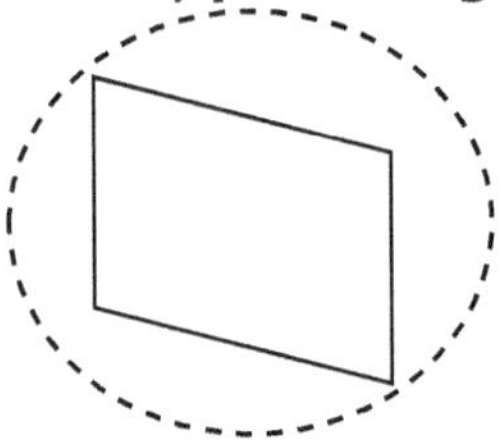

20. parallelogram
21. **trapezoid** 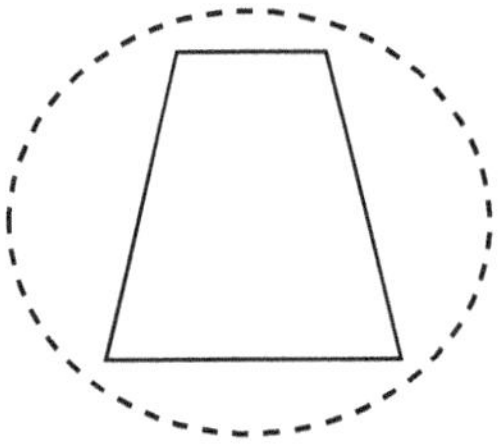
22. **diamond / rhombus** 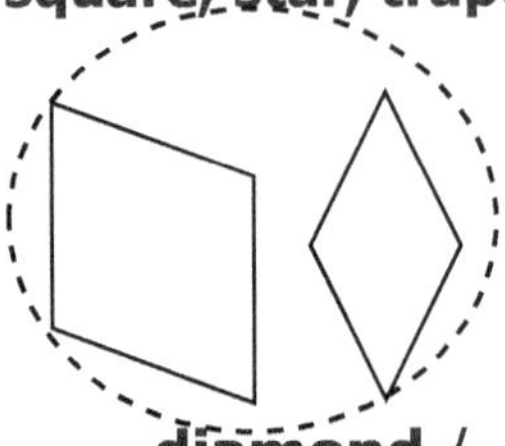
23. **octagon** 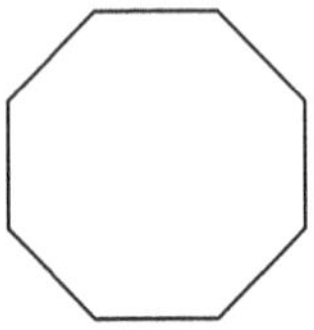
24. **triangle** 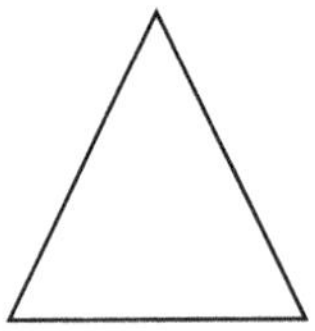
25. **hexagon** 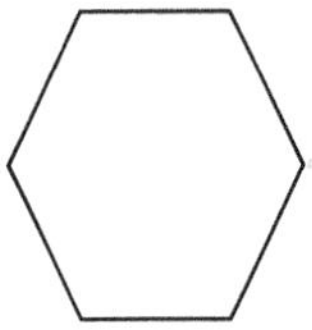
26. **cross** 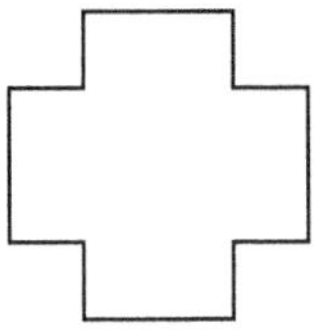
27. **pentagon** 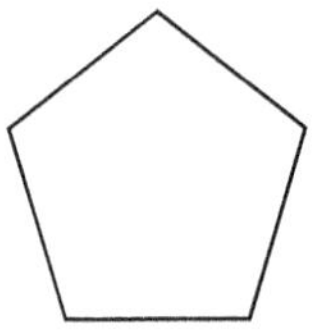
28. **circle** 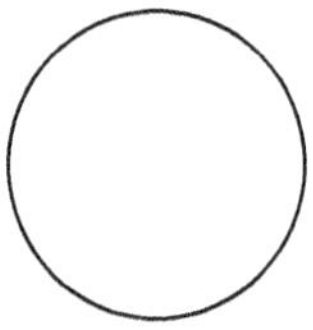
29. **decagon / star** 
30. **rectangle** 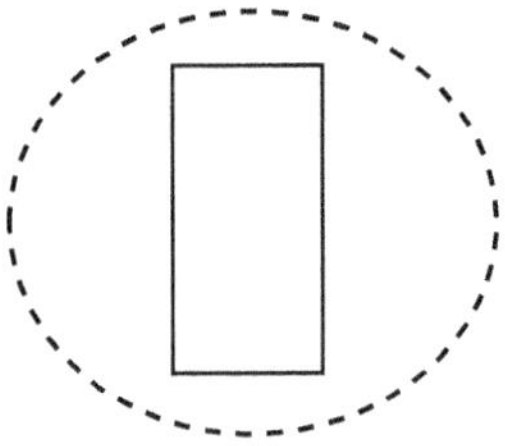
31. **oval** 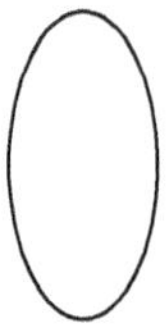
32. **heart** 
33. **heptagon** 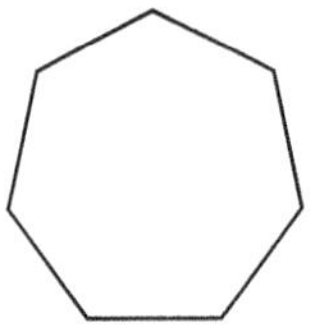
34. **nonagon** 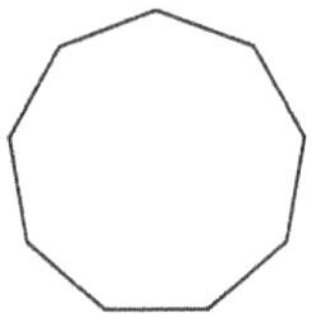
35. **square** 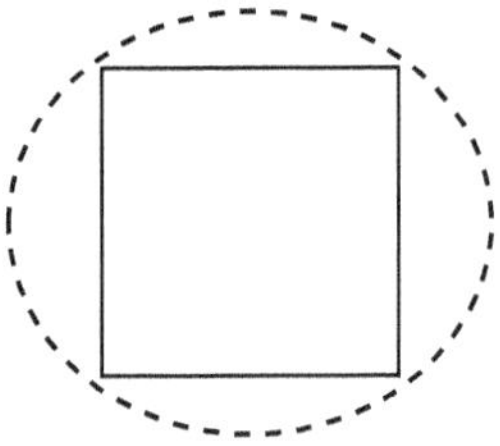

Directions: Write the name of each solid figure in the space provided.

cone, cube, cylinder, sphere

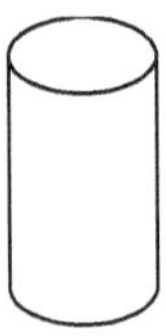

36. **cylinder**
37. **cone** 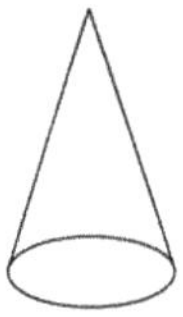
38. **sphere** 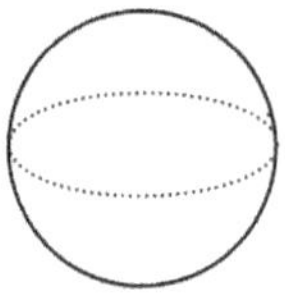
39. **cube**

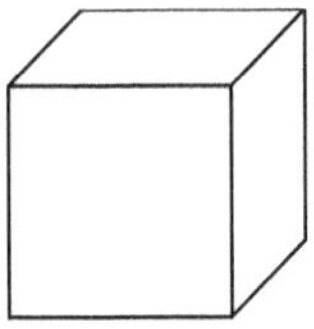

Name: _____   Date: _____

Key for Elementary Math Test #1
(Grades: kindergarten, 1st, 2nd, 3rd, 4th, 5th)

ODD AND EVEN NUMBERS

Directions: Categorize the numbers listed below.

Numbers: −10, 2, 9, 15

Odd		**Even**	
1.	**9**	3.	**2**
2.	**15**	4.	**−10**

5. Record any 2-digit <u>odd</u> number. **Odd numbers include 1, 3, 5, 7, or 9 or any number that ends with 1, 3, 5, 7, or 9, such as 4<u>3</u>.**

6. Record any 2-digit <u>even</u> number. **Even numbers include 0, 2, 4, 6, or 8 or any number that ends with 0, 2, 4, 6, or 8, such as 3<u>4</u>.**

NUMBER VALUES

7. How many arrows are in the set? Write the **cardinal** number in the space provided.

   8

8. What place is the underlined swimmer in? Write the **ordinal** number in the space provided.

   3rd

ROMAN NUMERALS

9. List the first 12 Roman numerals in order from 1 to 12.

   I, II, III, IV, V, VI, VII, VIII, IX, X, XI, XII

ADDITION AND SUBTRACTION FACTS

Directions: Solve.

10. 3 + 5 = **8**   12. 13 + 0 = **13**

11. 7 − 2 = **5**   13. 9 + 1 + 4 = **14**

Solve the equations. Show your work in the box.

14. $n - 7 = 9$ $n =$ **16** (Solution: 9 + 7 = 16)

15. $n + 5 = 15$ $n =$ **10** (Solution: 15 − 5 = 10)

Fill in the blanks. Use the words *difference, product, quotient,* or *sum*.

16. Addend + addend = **sum**

17. $n - n =$ **difference**

18. Factor × factor = **product**

19. Dividend ÷ divisor = **quotient**

Directions: Which pair of numbers equal the sum 8 when added together and the difference 4 when subtracted? Write the solution to the problem in the space provided.

20. Sum is 8
    Difference is 4

    6, 2   (6, 2   5, 3   7, 1)

PLACE VALUE

21. **2** hundreds   **1** tens   **4** ones

    Total number: **214**

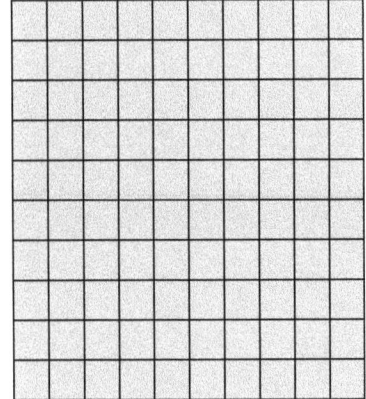  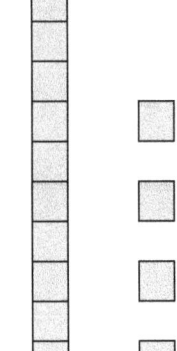

22. In the number 128, identify the number in the ones place, tens place, and hundreds place.

Ones	Tens	Hundreds
8	2	1

23. Fill in the blanks.

    6 hundreds + **8** tens + **2** ones = 682

Directions: Write the value of each underlined digit in the space provided.

      7̲92   79̲2   7̲92

24. **700**   25. **2**   26. **90**

27. In the spaces provided, rewrite each number in standard or expanded form.

Expanded Form						**Standard Form**
__400__	+	__10__	+	__7__	=	417

28. 600 + 50 + 4 = **654**

COUNTING FORWARDS
29. Count forwards from 1 to 100. Write each missing number in the space provided.

1	2	3	4	5	6	7	8	9	10
11	12	13	14	15	16	17	18	19	20
21	22	23	24	25	26	27	28	29	30
31	32	33	34	35	36	37	38	39	40
41	42	43	44	45	46	47	48	49	50
51	52	53	54	55	56	57	58	59	60
61	62	63	64	65	66	67	68	69	70
71	72	73	74	75	76	77	78	79	80
81	82	83	84	85	86	87	88	89	90
91	92	93	94	95	96	97	98	99	100

COUNTING BACKWARDS

30. Count backwards from 20 to 0. Write each missing number in the space provided.

20	19	**18**	**17**	**16**	**15**	**14**	**13**	**12**	**11**
10	**9**	**8**	**7**	**6**	**5**	**4**	**3**	**2**	**1**
0									

SKIP COUNTING

31. Skip count by 2's to 100. Write each missing number in the space provided.

2	4	**6**	**8**	**10**	**12**	**14**	**16**	**18**	**20**
22	**24**	**26**	**28**	**30**	**32**	**34**	**36**	**38**	**40**
42	**44**	**46**	**48**	**50**	**52**	**54**	**56**	**58**	**60**
62	**64**	**66**	**68**	**70**	**72**	**74**	**76**	**78**	**80**
82	**84**	**86**	**88**	**90**	**92**	**94**	**96**	**98**	**100**

32. Skip count by 5's to 100. Write each missing number in the space provided.

5	10	**15**	**20**	**25**	**30**	**35**	**40**	**45**	**50**
55	**60**	**65**	**70**	**75**	**80**	**85**	**90**	**95**	**100**

33. Skip count by 10's to 100. Write each missing number in the space provided.

10	20	**30**	**40**	**50**	**60**	**70**	**80**	**90**	**100**

34. Skip count by 25's to 100. Write each missing number in the space provided.

25	**50**	**75**	**100**

35. Skip count by 50's. Write the missing number in the space provided.

50	**100**

36. Skip count by 100's to 1,000. Write each missing number in the space provided.

100	200	**300**	**400**	**500**	**600**	**700**	**800**	**900**	**1000**

BEFORE, BETWEEN, AND AFTER NUMBERS (0-99)

37. In the spaces provided, write the missing before, between, and after numbers.

   <u>**7**</u> 8 9   31 <u>**32**</u> 33   57 58 <u>**59**</u>

MONEY

Directions: Write the <u>name</u> (such as *nickel*) and <u>value</u> (such as 25¢) of each coin.
dime, half-dollar, nickel, penny, quarter

| 38. **1¢** **penny**  | 39. **5¢** **nickel**  | 40. **10¢** **dime**  | 41. **25¢** **quarter**  | 42. **50¢** **half-dollar** |

Directions: Count the money and write the amount in the space provided. Use the symbols ¢ or $ in your answer.

| 43. **58¢**  | 44. **$1.96**  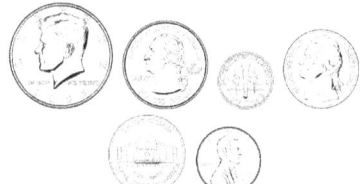 | 45. **$6.27**  |

46. Circle any combination of coins that are equivalent to .

Possible answers: 2 half-dollars; 4 quarters; 1 half dollar & 2 quarters; 1 half-dollar, 1 quarter, 2 dimes, & one nickel

MONEY ADDITION AND SUBTRACTION (INCLUDING 2-DIGIT MONEY ADDITION AND SUBTRACTION WITH/WITHOUT REGROUPING)

Directions: Complete each money addition and subtraction sentence. Line up the numbers vertically before adding or subtracting.

47. 71¢ + 6¢ =	48. 54¢ + 26¢ =	49. 89¢ − 6¢ =	50. 62¢ − 4¢ =
71¢ + 6¢ ――― **77¢**	¹54¢ + 26¢ ――― **80¢**	89¢ − 6¢ ――― **83¢**	⁵¹62¢ − 4¢ ――― **58¢**

Directions: Add the numbers (addends) with an addition sign. Subtract the numbers with a subtraction sign.

51.   62
    + 25
    ―――
    87

52.   ¹26
    + 8
    ―――
    34

53.   84
    − 20
    ―――
    64

54.   ⁶¹71
    − 43
    ―――
    28

TIME

Directions: What time is illustrated on the clock? When writing the time in digital form, remember to record the number for the little (hour) hand first; then write the number of minutes. If a second space is provided, use the words *o'clock, half past, quarter to*, or *quarter past* to write the time a different way.

55. **12:00**

   12 o'clock

56. **12:30**

   half past 12

57. **7:15**

    quarter past 7

58. **7:45**

    quarter to 8

59. **5:40**

60. How many seconds are in one minute? **60 seconds**

61. How many minutes are in one hour? **60 minutes**

62. How many hours are in one day? **24 hours**

FRACTIONS AND PERCENT

63. In the fractions below, is ¼ or ¾ a larger fraction? How do you know?

¾ (If you cut a whole pie into 4 equal slices, 3 slices would be greater than 1.)

Directions: What fraction of each circle is shaded? Write the fraction and use the initials **N** and **D** to mark the numerator and denominator.

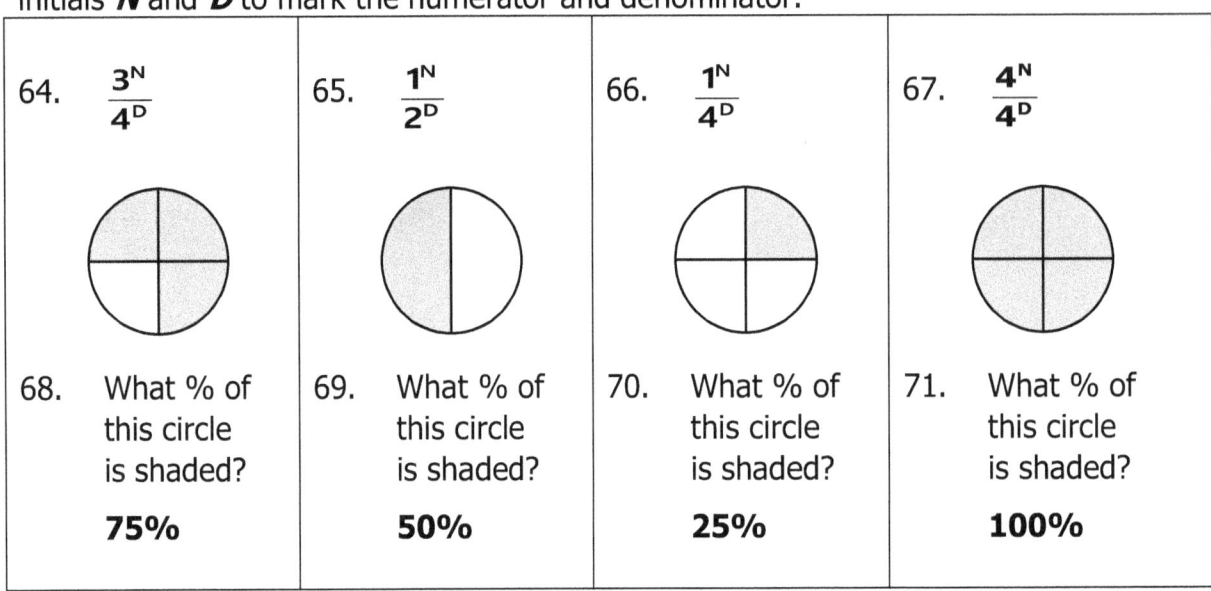

64. $\frac{3^N}{4^D}$

65. $\frac{1^N}{2^D}$

66. $\frac{1^N}{4^D}$

67. $\frac{4^N}{4^D}$

68. What % of this circle is shaded? **75%**

69. What % of this circle is shaded? **50%**

70. What % of this circle is shaded? **25%**

71. What % of this circle is shaded? **100%**

Name: _____ Date: _____

Key for Elementary Math Test #2
(Grades: kindergarten, 1st, 2nd, 3rd, 4th, 5th)

CALENDAR WORD PROBLEMS (DAYS OF A WEEK, DAY OF THE MONTH, MONTHS OF A YEAR, YEARS)

1. Write the abbreviation for each day of the week in the space provided.

   Abbreviations: Fri., Mon., Sat., Sun., Thurs., Tues., Wed.

Sunday	Monday	Tuesday	Wednesday	Thursday	Friday	Saturday
Sun.	**Mon.**	**Tues.**	**Wed.**	**Thurs.**	**Fri.**	**Sat.**

Directions: Read each sentence and use a calendar to help solve the problem. Write the day of the week in the space provided.

2. What day of the week is between Monday and Wednesday? **Tuesday**

3. What day of the week is before Friday? **Thursday**

4. What day of the week is after Friday? **Saturday**

5. If yesterday was Tuesday, what is today? **Wednesday**

6. If yesterday was Tuesday, what will tomorrow be? **Thursday**

7. If today is Thursday, what was yesterday? **Wednesday**

8. If today is Thursday, what will tomorrow be? **Friday**

9. If today is Monday, what day of the week will it be 4 days from now? **Friday**

10. If today is Friday, how many days are there until Sunday? **2**

Directions: Read each sentence and use a calendar to help solve the problem. Write the ordinal number in the space provided. (Examples of ordinal numbers are 1ˢᵗ, 2ⁿᵈ, 3ʳᵈ, 4ᵗʰ, and so on.)

11. What day of the month is between the 7ᵗʰ and 9ᵗʰ? **8ᵗʰ**

12. What day of the month is before the 29ᵗʰ? **28ᵗʰ**

13. What day of the month is after the 29ᵗʰ? **30ᵗʰ**

14. If yesterday was the 26ᵗʰ, what is today? **27ᵗʰ**

15. If yesterday was the 26ᵗʰ, what will tomorrow be? **28ᵗʰ**

16. If today is the 3ʳᵈ, what was yesterday? **2ⁿᵈ**

17. If today is the 3ʳᵈ, what will tomorrow be? **4ᵗʰ**

18. If today is the 6ᵗʰ, what date will it be 8 days from now? **14ᵗʰ**

19. If today is the 14ᵗʰ, what date will it be 1 week from now? **21ˢᵗ**

20. If today is the 5ᵗʰ, how many days are there until the 12ᵗʰ? **7**

21. Write the abbreviation for each month in the space provided.

    Abbreviations: Apr., Aug., Dec., Feb., Jan., Mar., Nov., Oct., Sept.

January	February	March	April	August	September	October	November	December
Jan.	**Feb.**	**Mar.**	**Apr.**	**Aug.**	**Sept.**	**Oct.**	**Nov.**	**Dec.**

Directions: Read each sentence and use the listed months to help solve the problem. Write the month in the space provided.

**January
February
March
April
May
June
July
August
September
October
November
December**

22. What month is between October and December? **November**

23. What month is before November? **October**

24. What month is after November? **December**

25. If the month is March, what will it be 2 months from now? **May**

Directions: Read each sentence and write the correct year in the space provided.

26. What year is between 2022 and 2024? **2023**

27. What year was just before 2021? **2020**

28. What year is just after 2021? **2022**

29. If the year is 2021, what will it be 3 years from now? **2024**

30. If the year is 2021, what was it 5 years ago? **2016**

Directions: Use the information given below to determine if Ashley, Aaron, or Jenna is the oldest.

31. Ashley is older than Jenna, but Aaron is older than Ashley. **Aaron**
    (Solution: Jenna   Ashley   Aaron)

ADDITION WORD PROBLEMS

Directions: Read each sentence and draw a picture to help solve the problem. Write the addition fact in the spaces provided.

32. The farmer had 0 pigs. He bought 8. How many pigs does the farmer have now?

    0 + 8 = 8

33. The farmer had 3 sheep. His ewe had 2 lambs. How many sheep does the farmer have now?

    3 + 2 = 5

34. This morning, Alex milked 3 cows. Kailey milked 5. How many cows were milked altogether?

    3 + 5 = 8

35. There were 8 chicks; 3 more hatched from their eggs. How many chicks in all?

    8 + 3 = 11

36. Caleb caught 7 ducks. Jacob caught 3. How many ducks were caught altogether?

    7 + 3 = 10

37. Ryan pulled up 10 carrots. Nathan pulled up 8. How many carrots were pulled up altogether?

    10 + 8 = 18

38. Mandi peeled 5 potatoes. Taylor peeled 2. How many potatoes were peeled altogether?

    5 + 2 = 7

39. Jessica picked 7 tomatoes from the vine. Joshua picked 9. How many tomatoes were picked altogether?

    7 + 9 = 16

SUBTRACTION WORD PROBLEMS

Directions: Read each sentence and draw a picture to help solve the problem. Write the subtraction fact in the spaces provided.

40. The farmer has 9 pigs; 0 have been fed this morning. How many pigs need to be fed?

    9 – 0 = 9

41. The farmer had 7 sheep. He sold 2 ewes. How many sheep does the farmer have left?

    7 – 2 = 5

42. Sam has 5 cows to milk. Spencer milked 4 for him. How many cows still need to be milked?

    5 – 4 = 1

43. Beth's class has been waiting for 8 chicks to hatch. This morning, they found that 4 hatched overnight. How many chicks are they still waiting for?

    8 – 4 = 4

44. David and Michael saw 6 ducks. They caught 2. How many ducks got away?

    6 – 2 = 4

45. I had 8 carrots. I gave Grace 1. How many carrots do I have left?

    8 – 1 = 7

46. There were 9 potatoes growing in the ground. The farmer dug up 4 to take to market. How many potatoes are still in the ground?

    9 – 4 = 5

47. There were 16 tomatoes growing on a vine. Eliana picked 5. How many tomatoes are left on the vine?

    16 – 5 = 11

48. There were 19 ears of corn on the stalk. The farmer picked 10 to take to market. How many ears of corn are left on the stalk?

    19 – 10 = 9

ADDITION AND SUBTRACTION WORD PROBLEMS

Directions: Read each sentence and draw a picture to help solve the problem. Write the addition or subtraction fact in the spaces provided.

49. There were 15 slices of ham. If 7 slices were eaten, how many slices are left?

    15 − 7 = 8

50. Sophia has 1 red apple. Liam has 1 green apple. How many apples do they have altogether?

    1 + 1 = 2

51. Noah caught 2 fish. His brother caught 7. How many fish did they catch altogether?

    2 + 7 = 9

52. My aunt made 6 quilts. She sold 4. How many quilts does she have left?

    6 − 4 = 2

53. Our football team has played 9 games so far. We lost 6 games. How many games did we win?

    9 − 6 = 3

54. Grandma baked 4 pumpkin pies, and Mama baked 4 peach pies. How many pies were baked altogether?

    4 + 4 = 8

55. Emma found 11 pecans. She fed 7 to the squirrels. How many pecans does Emma have left?

    11 − 7 = 4

56. Lucas raked 6 bags of leaves. His dad raked up 1 more bag. How many bags of leaves in all?

    6 + 1 = 7

57. I gathered 13 eggs from the barn. I accidentally broke 6. How many eggs do I have left?

    13 − 6 = 7

MONEY ADDITION WORD PROBLEMS

58. Hailee had five nickels. She found 60¢ on her way to the store. How much money does she have now?

    25 + 60 = **85¢**

59. Judah has three dimes, two nickels, and six pennies. How much money does he have in all?

    30 + 10 + 6 = **46¢**

60. Elizabeth had four dimes. She earned 50¢ more. How much money does Elizabeth have now?

    40 + 50 = **90¢**

61. Emily had one quarter. Her mom gave her $3.00 more. How much money does she have now?

    .25 + $3.00 = **$3.25**

62. Bret wants to buy a set of baseball cards. It costs 96¢. He has 2 quarters and 8 nickels. Does he have enough money? (Show your work.)

    25 + 25 + 5 + 5 + 5 + 5 + 5 + 5 + 5 + 5 = **90¢, no**

63. Amishia wants to buy some candy. The cost is 84¢. She has 1 half-dollar, 3 dimes, and 1 penny. Does she have enough money? (Show your work.)

    50 + 10 + 10 + 10 + 1 = **81¢, no**

MONEY SUBTRACTION WORD PROBLEMS

64. Miranda had 48¢. She lost two nickels. How much money does she have left?

    48 − 10 = **38¢**

65. Jimmy has two dimes, four nickels, and five pennies. If he saves 45¢, how much can he spend?

    20 + 20 + 5 = 45; 45 − 45 = **0¢**

66. Kim had 43¢. She spent two dimes at the store. How much money does she have left?

    43 − 20 = **23¢**

67. Sophie had $1.85. She gave her sister three quarters. How much money does she have left?

    $1.85 − .75 = **$1.10**

MONEY ADDITION AND SUBTRACTION WORD PROBLEMS

68. Mia had $1.25. She gave her brother one quarter. How much money does she have left?

    $1.25 − .25 = **$1.00**

69. Ava had two quarters. Her mom gave her $3.00 more. How much money does she have now?

    .50 + $3.00 = **$3.50**

70. Elijah had 88¢. He lost five nickels. How much money does he have left?

    88 − 25 = **63¢**

71. Logan has three dimes, four nickels, and one penny. If he saves 40¢, how much can he spend?

    30 + 20 + 1 = 51; 51 − 40 = **11¢**

72. Olivia has four dimes, one nickel, and four pennies. How much money does she have in all?

    40 + 5 + 4 = **49¢**

73. Oliver had three dimes. He earned 10¢ more. How much money does Oliver have now?

    30 + 10 = **40¢**

74. Amelia had 84¢. She spent four dimes at the store. How much money does she have left?

    84 − 40 = **44¢**

Directions: Circle the correct change. Add the change and write the amount.

75. Jodi spent 68¢ at the store. She gave the store cashier $1.00. How much change did she receive?

    $1.00 − .68 = **32¢**

76. Derek spent $1.37 at the store. He gave the store clerk $5.00. How much change did he receive?

    $5.00 − $1.37 = **$3.63**

77. Kingston had $20.00. He spent $12.24 at the book fair. How much change did he receive?

    $20.00 − $12.24 = **$7.76**

92

TIME WORD PROBLEMS

Directions: Read each sentence and use a demonstration clock (and the time zone chart when necessary) to help solve the problem. Write the time in the space provided.

78. The farmer usually starts his chores at 6:30 A.M. Today he started them 30 minutes late. What time did he start his chores?

    7:00 A.M.

79. School starts at 8:30 A.M. Today, Shana was 15 minutes early. What time did she get to school?

    8:15 A.M.

80. Hunter usually eats lunch at 1:00 P.M. Today he ate it 45 minutes late. What time did he eat lunch?

    1:45 P.M.

81. The farmer usually goes to bed at 10:30 P.M. Today he went to bed 1 hour early. What time did he go to bed?

    9:30 P.M.

82. If it is 10:45 P.M., what time will it be 5 minutes later?

    10:50 P.M.

TIME ZONES					
Hawaii	Alaska	Pacific	Mountain	Central	Eastern
12:00	1:00	2:00	3:00	4:00	5:00

83. If it's 3:00 P.M. in Hawaii, what time is it in Colorado (Mountain time)?

    6:00 P.M. (3 hours later)

84. If it's 7:30 A.M. in Alaska, what time is it in Hawaii? **6:30 A.M.** (1 hour earlier)

85. If it's 1:00 P.M. in California (Pacific time), what time is it in Kansas (Central time)?

    3:00 P.M. (2 hours later)

86. If it's 2:30 A.M. in New Mexico (Mountain time), what time is it in Washington (Pacific time)?

    1:30 A.M. (1 hour earlier)

87. If it's 7:00 P.M. in Arkansas (Central time), what time is it in New York (Eastern time)?

    8:00 P.M. (1 hour later)

88. If it's 12:30 A.M. in Florida (Eastern time), what time is it in Utah (Mountain time)?

    10:30 P.M. (2 hours earlier)

89. If it's 5:00 P.M. in Alabama (Central time), what time is it in Tennessee (Central time)?

    5:00 P.M. (The same time)

FRACTION WORD PROBLEMS

Directions: Read each sentence and draw a picture to help solve the problem. Write the fraction (and the % if asked for) in the space provided.

90. There were 4 pieces of apple pie. I ate 1 slice. What percentage of the pie did I eat?

    ¼ = 25%

91. There were 2 pieces of peach pie. I ate 1 slice. What percentage of the pie did I eat?

    ½ = 50%

92. There were 4 pieces of pecan pie. We ate 3 slices. What percentage of the pie did we eat?

    ¾ = 75%

93. There were 7 pieces of pumpkin pie. We ate 3 slices. How much of the pie did we eat?

    $$\frac{3}{7}$$

Name: _____   Date: _____

Key for Elementary Math Test #3
(Grades: kindergarten, 1st, 2nd, 3rd, 4th, 5th)

Supplies: Ruler

PATTERNS (Shapes & Numbers)
1. Assign a capital letter to each shape. Begin with the letter *A* and assign the same letter to each matching shape. Then continue the pattern in the boxes.

    ★ ● ■ ■ ★ ● ■ ■ | ★ | ● | ■ | ■ |
    A B C C A B C C | **A** | **B** | **C** | **C** |

2. What would the next number of the pattern be?
    10, 15, 20, 25…

    30 (Solution: 25 + 5 = 30)

GREATER- AND LESS-THAN NUMBERS
3. Write **>**, **<**, or a greater- or less-than number in the space provided.
    Any number less than 17 can be used for the middle answer.
    2 **<** 5   _____ < 17   89 **>** 53

ORDERING NUMBERS FROM LEAST TO GREATEST (0-99)
4. In the spaces provided, write the numbers in order from least to greatest.
    24   6   19

    6   **19**   **24**

ROUNDING NUMBERS TO THE NEAREST TEN AND HUNDRED
5. Round the numbers to the nearest ten.

Number	Rounded
65	**70**
17	**20**

6. Round the numbers to the nearest hundred.

Number	Rounded
619	**600**
350	**400**

7. According to the bar graph below, in what year did Elizabeth read the most number of books?

   2021

8. According to the line graph, in what year did Shana read the least number of books?

   2017

9. According to the bar graph, in what year did Elizabeth read 7 books?

   2019

10. According to the line graph, the number of books Shana read decreased between what two years?

    2018-2019

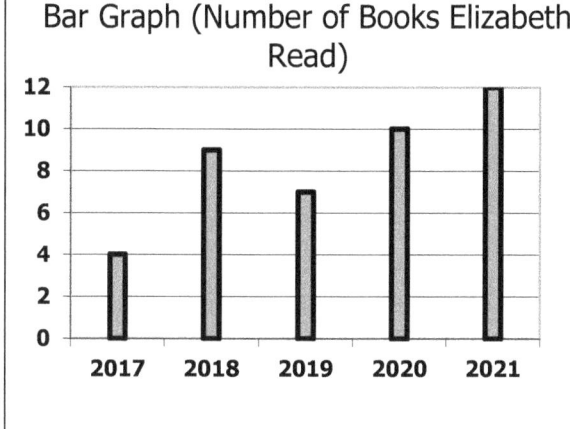
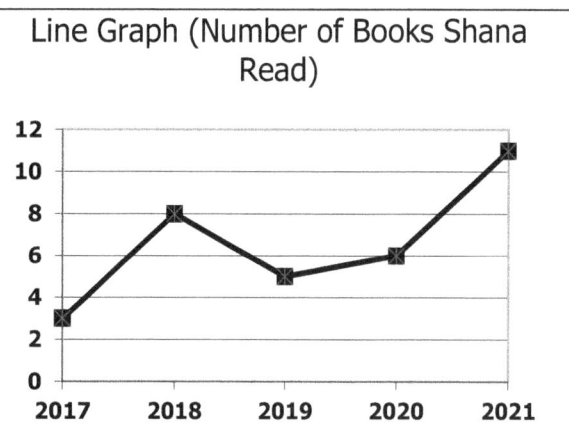

11. Use a ruler to make a line graph for the data below.

Year	Number of animals on Uncle Jim's farm
1995	8
2000	15
2005	21
2010	18
2015	25

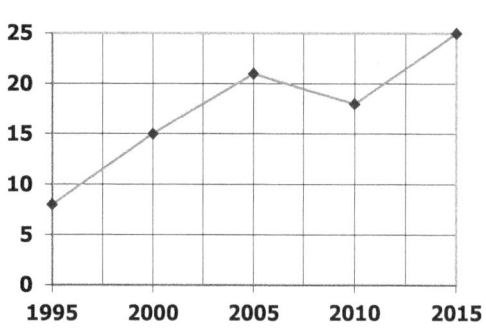

Find the perimeter.

12. Perimeter

    18

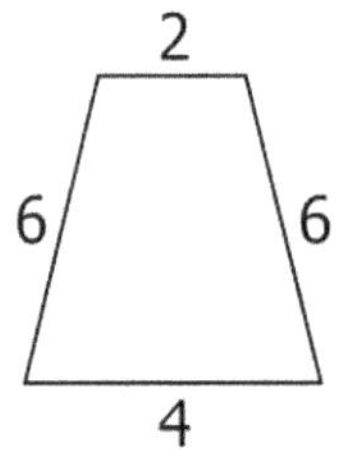

13. What is the area of the square to the right?

    9 square units

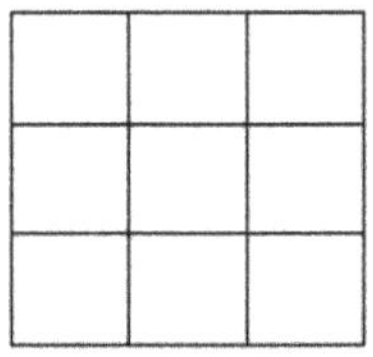

GEOMETRY

14. Circle the shape below that is **similar** to the shape in the box.

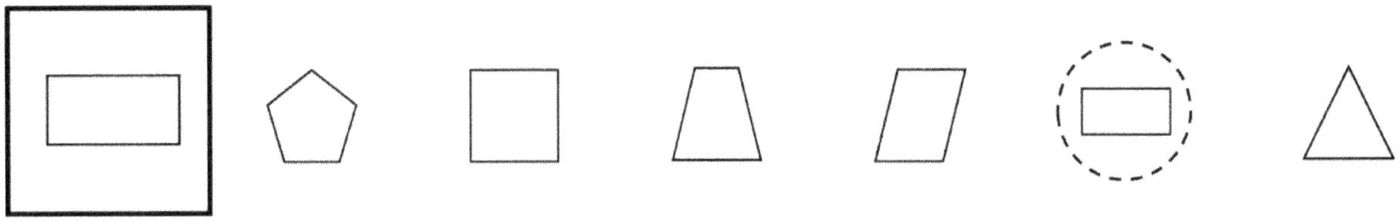

15. Circle the shape below that is **congruent** to the shape in the box.

16. Use your ruler to show how many axes of symmetry (up to two) each figure has. Although some of the shapes have more than two lines of symmetry, illustrate the one or two most obvious ones. Note that one of the shapes doesn't have any lines of symmetry.

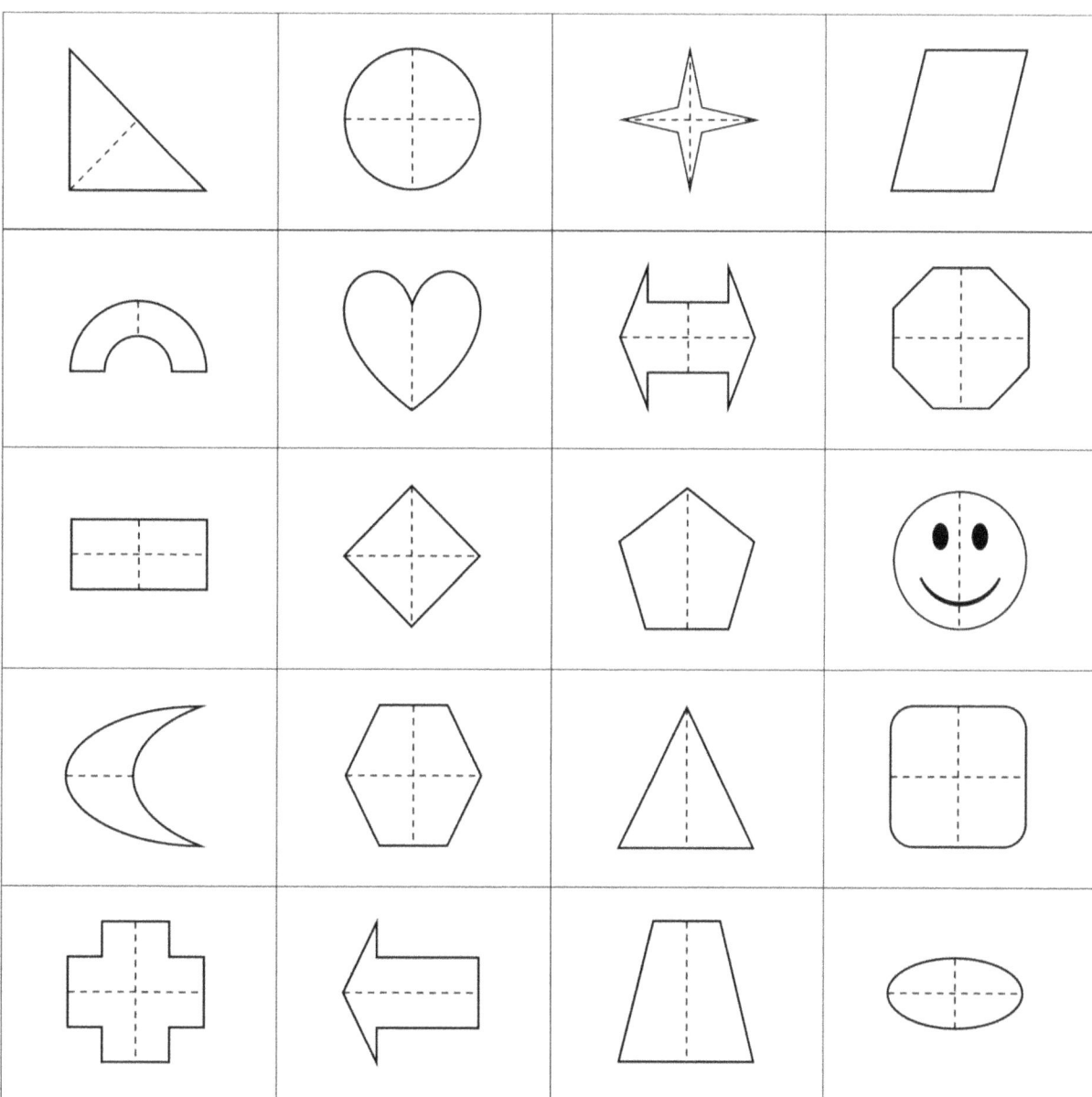

TEMPERATURE

Directions: What temperature is illustrated on the thermometer? Write the temperature in the space provided. Then circle the word that describes the temperature (use the guidelines below to help you).
  Freezing: 32° or lower
  Cold: 33°-63°
  Warm: 64°-84°
  Hot: 85° or higher

17. **90°**   freezing   cold   warm   (hot)

Directions: Read each sentence and use the thermometer to help solve the problem. (The thermometer does not illustrate the temperature from the word problems.) Write the temperature in the space provided.

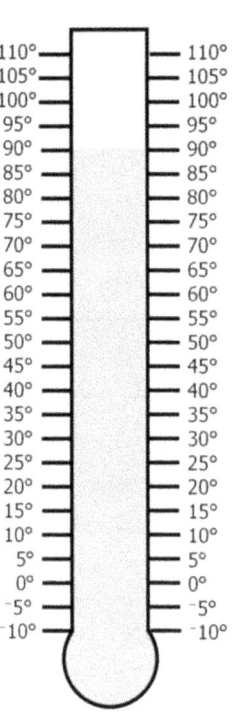

18. At 8:00 this morning, the temperature was 75°. By noon, it had risen 20°. What was the temperature at noon?

    95°

19. At 5:00 this afternoon, the temperature was 85°. By 9:00 P.M., it had dropped 5°. What was the temperature at 9:00 P.M.?

    80°

MEASUREMENT (LENGTH)

Directions: Write the measurement in the space provided (*1 foot*, *1 inch*, or *1 yard*).

20. 25 millimeters (mm) ≈ **1 inch**

21. 2½ centimeters (cm) ≈ **1 inch**

22. 12 inches (in.) = **1 foot**

23. 3 feet (ft.) = **1 yard**

24.	Use a ruler to measure the line in the box to the nearest inch. **3 inches** _____
25.	Use a ruler to draw 25 millimeters in the box.
26.	Use a ruler to draw one centimeter in the box.
27.	Use a ruler to draw 1 inch in the box.

THIS MARKS THE END OF THE TEST FOR GRADES KINDERGARTEN THROUGH 2ND. STUDENTS IN GRADES 3RD THROUGH 5TH SHOULD CONTINUE.

CONVERSION PROBLEMS

Directions: Solve the conversion problems.

28. 6 feet = **72** inches (Solution: 12 in = 1 ft; 6 × 12 = 72)

29. 24 in = **2** ft (Solution: 12 in = 1 ft; 24 ÷ 12 = 2)

30. 18 feet = **6** yards (Solution: 3 ft = 1 yd; 18 ÷ 3 = 6)

31. 2 yd = **6** ft (Solution: 3 ft = 1 yd; 2 × 3 = 6)

32. How many seconds are in six minutes?
    6 minutes = **360** seconds (Solution: 60 sec = 1 min; 60 × 6 = 360)

33. How many minutes are in 180 seconds?
    180 sec = **3** min (Solution: 60 sec = 1 min; 180 ÷ 60 = 3)

34. How many hours are in 300 minutes?
    300 minutes = **5** hours (Solution: 60 min = 1 hr; 300 ÷ 60 = 5)

35. How many minutes are in six hours?
    6 hr = **360** min (Solution: 60 min = 1 hr; 60 × 6 = 360)

36. How many hours are in six days?

    6 days = **144** hours (Solution: 24 hr = 1 day; 24 × 6 = 144)

37. How many days are in 48 hours?

    48 hours = **2** days (Solution: 24 hr = 1 day; 48 ÷ 24 = 2)

MULTIPLICATION AND DIVISION FACTS

SKIP COUNTING

38. Skip count by 2's to 18. Write each missing number in the space provided.

    2   **4**   **6**   **8**   **10**   **12**   **14**   **16**   **18**

39. Skip count by 3's to 27. Write each missing number in the space provided.

    3   **6**   **9**   **12**   **15**   **18**   **21**   **24**   **27**

40. Skip count by 4's to 36. Write each missing number in the space provided.

    4   **8**   **12**   **16**   **20**   **24**   **28**   **32**   **36**

41. Skip count by 5's to 45. Write each missing number in the space provided.

    5   **10**   **15**   **20**   **25**   **30**   **35**   **40**   **45**

42. Skip count by 6's to 54. Write each missing number in the space provided.

    6   **12**   **18**   **24**   **30**   **36**   **42**   **48**   **54**

43. Skip count by 7's to 63. Write each missing number in the space provided.

    7   **14**   **21**   **28**   **35**   **42**   **49**   **56**   **63**

44. Skip count by 8's to 72. Write each missing number in the space provided.

    8   **16**   **24**   **32**   **40**   **48**   **56**   **64**   **72**

45. Skip count by 9's to 81. Write each missing number in the space provided.

$\underline{\quad 9 \quad}$ $\underline{\quad 18 \quad}$ $\underline{\quad 27 \quad}$ $\underline{\quad 36 \quad}$ $\underline{\quad 45 \quad}$ $\underline{\quad 54 \quad}$ $\underline{\quad 63 \quad}$ $\underline{\quad 72 \quad}$ $\underline{\quad 81 \quad}$

Directions: Solve.

46. 8 × 2 = **16**

47. 5 × 1 = **5**

48. 7 × 0 = **0**

49. 18 ÷ 6 = **3**

50. 4 ÷ 1 = **4**

Directions: What does 2 times 3 equal (2 × 3 = __). Use stars or other shapes and the 2 cells below to illustrate the answer.

51. 2 × 3 = **6**

☆☆ ☆	☆☆ ☆

Directions: What does 8 divided by 2 equal (8 ÷ 2 = __). Divide the 8 stars between the 2 cells below to illustrate the answer. You can cross out the stars as you use them.

52. 8 ÷ 2 = **4**

☆ ☆ ☆ ☆ ☆ ☆ ☆ ☆

☆☆ ☆☆	☆☆ ☆☆

MULTIPLICATION WORD PROBLEMS

Directions: Read each sentence and draw a picture to help solve the problem. Write the multiplication fact in the spaces provided.

53. Each of the 15 farmers have 0 white pigs. How many white pigs do they have in all?

    15 × 0 = 0

54. Each of the farmer's 4 ewes had 2 lambs. How many lambs in all?

    4 × 2 = 8

55. This morning, 5 children milked 1 cow apiece. How many cows were milked altogether?

    5 × 1 = 5

56. Each of the 4 mother hens have 7 eggs that have started to hatch. How many chicks will there be in all?

    4 × 7 = 28

57. Three children caught 5 ducks apiece. How many ducks were caught altogether?

    3 × 5 = 15

58. Seven children pulled up 7 carrots apiece. How many carrots were pulled up altogether?

    7 × 7 = 49

59. Nine children peeled 6 potatoes apiece. How many potatoes were peeled altogether?

    9 × 6 = 54

60. Two children picked 5 tomatoes from the vine. How many tomatoes were picked altogether?

    2 × 5 = 10

DIVISION WORD PROBLEMS

Directions: Read each sentence and draw a picture to help solve the problem. Write the division fact in the spaces provided.

61. There were 28 carrots; 7 children divided them. How many carrots did each child get?

    28 ÷ 7 = 4

62. There were 12 celery sticks. Cara and Pam divided them. How many sticks did each one get?

    12 ÷ 2 = 6

63. Six potatoes need to be peeled; 2 children divided them. How many potatoes did each child peel?

    6 ÷ 2 = 3

64. Eighteen tomatoes were picked from the vine; 3 children divided them. How many tomatoes did each child get?

    18 ÷ 3 = 6

MULTIPLICATION AND DIVISION WORD PROBLEMS

Directions: Read each sentence and draw a picture to help solve the problem. Write the multiplication or division fact in the spaces provided.

65. There were 21 slices of ham. If 7 people divided them, how many slices did each person get?

    21 ÷ 7 = 3

66. There were 24 red apples. Eight children divided them. How many apples did each child get?

    24 ÷ 8 = 3

67. Five women were selling quilts at the flea market. If each woman sold 6 quilts, how many quilts were sold altogether?

    5 × 6 = 30

68. Nine children were fishing. If each child caught 4 fish, how many fish were caught altogether?

    9 × 4 = 36

69. There were 5 slices of pumpkin pie. Five people divided them. How many slices did each person get?

    5 ÷ 5 = 1

70. Two children gathered 2 eggs apiece from the barn. How many eggs were gathered altogether?

    2 × 2 = 4

71. Four brothers found 32 turkey feathers in their yard. If they divide them evenly, how many feathers will each brother get?

    32 ÷ 4 = 8

72. Seven children were raking leaves. If each child raked up 5 bags full, how many bags of leaves were there?

    7 × 5 = 35

73. Abigail had 30 pecans. She divided them between 5 squirrels. How many pecans did each squirrel get?

    30 ÷ 5 = 6

Name: _____   Date: _____

Key for Elementary Math Test #4
(Grades: 3rd, 4th, 5th)

Directions: Show all your work in the boxes provided.

1. In an ordered pair such as (2, 4), 2 represents the **x**-coordinate, and 4 represents the **y**-coordinate. (Fill in the blanks with *x* or *y*.)

Directions: Graph the following ordered pairs in the coordinate plane below.

2. A (7, −7)

3. B (−1, 0)

4. C (0, 7)

5. D (−5.5, 2½)

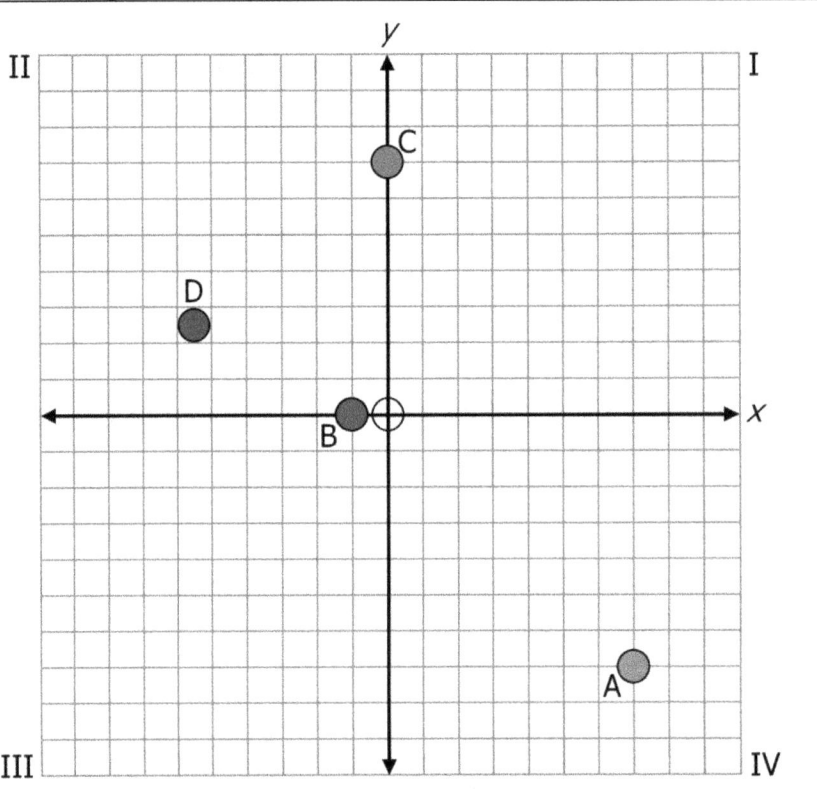

6. Circle the origin in the coordinate plane above.

7. The four sections that a coordinate plane is divided into by the *x*- and the *y*-axes are called **quadrants**. These are numbered I, II, III, and IV.

108

8. Sketch a standard (x, y) coordinate plane in the box to the right and use it to find the location of a square's corner. (The three known corners of the square are listed below.) Record the point of the fourth corner in the space provided.
(−3, −3), (3, 3), (3, −3)

(−3, 3)

Addition and Subtraction (with regrouping)

Directions: Rewrite the problems vertically to solve.

9. 236 + 45 = **281**
 	
    ¹
  236
 + 45
  281
	10. 6943 + 728 = **7,671**
	
	  1 1
	  6943
	 + 728
	  7671

11. 72 − 54 = **18**
 	
  6 1
  7̸2
 − 54
   18
	12. 403 − 35 = **368**
	
	  3 9 1
	  4̸0̸3
	 − 35
	  368

Multiplication (with and without regrouping)

Directions: Rewrite the problems vertically to solve.

13. 63 × 4 = **252**
 	
  ¹63
 × 4
  252
	14. 32 × 46 = **1,472**
	
	   ¹32
	 × 46
	  ¹192
	  1280
	  1472

15. 421 × 213 = **89,673**

```
   421
 × 213
  1263
  4210
 84200
 89673
```

16. 43 × 20 = **860**

```
   43
 × 20
  860
```

17. 241 × 302 = **72,782**

```
  ¹241
 × 302
   482
 72300
 72782
```

True or False

18. You can change a remainder to a fraction by writing the remainder over the divisor.

    True

19. 37 ÷ 5 = **7 r 2**

20. Rewrite the remainder in 37 ÷ 5 as a fraction. **$\frac{2}{5}$**

Long Division

21. 72 ÷ 2 = **36**

```
    36
 2)72
    6
   12
```

22. 890 ÷ 5 = **178**

```
   178
5)890
   5
   39
   35
   40
```

23. 195 ÷ 15 = **13**

```
    13
15)195
    15
    45
```

Division (0 Remaining)

Directions: Turn the remainder into a decimal.

24. 7 ÷ 4 = **1.75**

```
   1.75
4)7.00
   4.
   3.0
   2.8
    20
```

Directions: Divide the decimals.

Division (Decimal Dividend)

25. 47.1 ÷ 3 = **15.7**

```
    15.7
3)47.1
    3.
    17
    15
    21
```

Decimal Dividend and Divisor

26. 6.32 ÷ 0.4 = **15.8**

```
      15.8
0.4)6.32
      4.
      23
      20
      32
```

Decimal Dividend and Divisor

27. 0.186 ÷ 0.02 = **9.3**

```
         9.3
0.02)0.186
        18
         6
         6
```

Decimals in Addition and Subtraction Problems

Directions: Line up the decimals to solve.

28.   7.2 + 61.3 = **68.5**
 	
         7.2
       +61.3
        68.5
	29. 18.69 − 5.24 = **13.45**
	
	       18.69
	      − 5.24
	       13.45

30.   9.5 − .35 = **9.15**
 	
       4 1
      9.5̶0
     − .35
      9.15
	31. 679.1 − 85 = **594.1**
	
	     5 1
	    6̶79.1
	    − 85.0
	     594.1

Decimals in Multiplication

Directions: Rewrite the problems vertically to solve.

32.   7.4 × 0.2 = **1.48**
 	
       7.4
     × 0.2
      1.48
	33. 0.15 × 0.3 = **0.045**
	
	      1
	     0.15
	    × 0.3
	    0.045

34. 1.742 × 0.5 = **0.871**	3 2 1
	     1.742
	    × 0.5
	     .8710

35. On the top half of the chart below, fill in the missing word name. **tenths**

36. On the bottom half of the chart below, fill in the missing number. $\dfrac{1}{10}$

thousands	hundreds	tens	ones	tenths	hundredths	thousandths	ten thousandths
1,000	100	10	1	$\dfrac{1}{10}$	$\dfrac{1}{100}$	$\dfrac{1}{1,000}$	$\dfrac{1}{10,000}$

Directions: Rewrite each expanded numeral as a decimal.

37. $5(1,000) + 8(100) + 4(10) + 9(1) + 6(\dfrac{1}{10}) + 3(\dfrac{1}{100}) + 2(\dfrac{1}{1,000})$

    5849.632

38. $8(1,000) + 3(100) + 9(10) + 5(\dfrac{1}{10}) + 7(\dfrac{1}{100}) + 4(\dfrac{1}{1,000}) + 6(\dfrac{1}{10,000})$

    8390.5746

39. $5(\dfrac{1}{10,000}) + 4(100) + 7(1,000) + 2(\dfrac{1}{100}) + 6(\dfrac{1}{10}) + 3(10) + 1(\dfrac{1}{1,000})$

    7430.6215

Directions: Rewrite each word number as a decimal.

40. Three hundred forty-two thousandths   **0.342**

41. Six hundred twenty-seven ten thousandths   **0.0627**

42. Four thousand seven hundred nineteen and six hundred fifty-two thousandths

    4,719.652

Directions: Use the symbol <, >, or = to compare the decimals below.

43. 7.2043 **<** 7.206

44. 5.2 **=** 5.20

Directions: What fraction of each circle is shaded? Write the fraction in the space provided. (If the fraction can be reduced, write it after the equal sign.) Then use the initials **N** and **D** to mark the numerator and denominator.

45. $\dfrac{2^N}{8^D} = \dfrac{1}{4}$	46. $\dfrac{5^N}{10^D} = \dfrac{1}{2}$	47. $\dfrac{9^N}{12^D} = \dfrac{3}{4}$
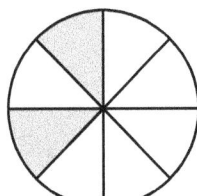	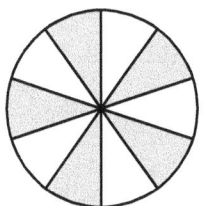	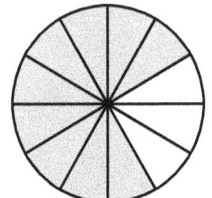
48. What percent of this circle is shaded? **25%**	49. What percent of this circle is shaded? **50%**	50. What percent of this circle is shaded? **75%**

51. In the fraction ¾, 3 is the **numerator**, and 4 is the **denominator**.

Proper and Improper Fractions

52. Give an example of a proper fraction.

The top number must be less than the bottom, such as $\dfrac{3}{4}$.

53. Give an example of an improper fraction.

The top number must be greater than the bottom, such as $\dfrac{4}{3}$.

Directions: Rewrite the fractions in lowest terms or as integers.

54. $\dfrac{0}{8} = \mathbf{0}$

55. $\dfrac{9}{9} = \mathbf{1}$

56. $\dfrac{5}{-5} = \mathbf{-1}$

57. $\dfrac{3}{1} = \mathbf{3}$

58. $\dfrac{32}{4} = \mathbf{8}$

59. $\dfrac{6}{21} = \dfrac{2}{7}$

Directions: Rewrite the fraction as a mixed number. Reduce to lowest terms.

60. $\dfrac{36}{10} = 3\dfrac{6}{10} = 3\dfrac{3}{5}$

Directions: Rewrite the mixed number as an improper fraction.

61. $5\dfrac{3}{4} = \dfrac{23}{4}$

Adding and Subtracting Fractions (Like Denominators)

Directions: Identify the sum or difference. (Always remember to reduce fractions to lowest terms and to rewrite improper fractions as mixed numbers.)

62. $\dfrac{5}{9} + \dfrac{4}{9} = \dfrac{9}{9} = 1$

63. $\dfrac{6}{8} - \dfrac{2}{8} = \dfrac{4}{8} = \dfrac{1}{2}$

64. $\dfrac{12}{6} - \dfrac{4}{6} = \dfrac{8}{6} = 1\dfrac{2}{6} = 1\dfrac{1}{3}$

65. $\dfrac{6}{17} + \dfrac{1}{17} + \dfrac{4}{17} = \dfrac{11}{17}$

Adding and Subtracting Fractions (Different Denominators)

Directions: When adding or subtracting fractions with different denominators, rewrite the problems vertically to solve. After identifying the sum or difference, remember to reduce fractions to lowest terms and to rewrite improper fractions as mixed numbers.

66. $\dfrac{4}{12} + \dfrac{1}{6} = \dfrac{6}{12} = \dfrac{1}{2}$

$$\begin{aligned}\dfrac{4}{12} &= \dfrac{4}{12}\\+\dfrac{1}{6} &= \dfrac{2}{12}\\\hline\dfrac{6}{12} &= \dfrac{1}{2}\end{aligned}$$

67. $\dfrac{3}{6} - \dfrac{2}{9} = \dfrac{5}{18}$

$$\begin{aligned}\dfrac{3}{6} &= \dfrac{9}{18}\\-\dfrac{2}{9} &= \dfrac{4}{18}\\\hline&\dfrac{5}{18}\end{aligned}$$

68. $$\begin{aligned}\dfrac{2}{10} &= \dfrac{2}{10}\\-\dfrac{3}{5} &= \dfrac{6}{10}\\\hline-\dfrac{4}{10} &= -\dfrac{2}{5}\end{aligned}$$

69. $\dfrac{3}{4} + \dfrac{5}{8} + \dfrac{1}{2} = 1\dfrac{7}{8}$

$$\begin{aligned}\dfrac{3}{4} &= \dfrac{6}{8}\\\dfrac{5}{8} &= \dfrac{5}{8}\\+\dfrac{1}{2} &= \dfrac{4}{8}\\\hline\dfrac{15}{8} &= 1\dfrac{7}{8}\end{aligned}$$

Adding and Subtracting Mixed Numbers (Like Denominators)

Directions: Add or subtract the mixed numbers. Reduce to lowest terms.

70.	$2\frac{4}{7} + 3\frac{2}{7} = \mathbf{5\frac{6}{7}}$ (Solution: $\frac{18}{7} + \frac{23}{7} = \frac{41}{7} = 5\frac{6}{7}$)
71.	$4\frac{5}{8} - 1\frac{3}{8} = \mathbf{3\frac{2}{8} = 3\frac{1}{4}}$ (Solution: $\frac{37}{8} - \frac{11}{8} = \frac{26}{8} = 3\frac{2}{8} = 3\frac{1}{4}$)
72.	$5\frac{6}{8} + 2\frac{3}{8} = \mathbf{8\frac{1}{8}}$ (Solution: $\frac{46}{8} + \frac{19}{8} = \frac{65}{8} = 8\frac{1}{8}$)

Adding and Subtracting Mixed Numbers (Different Denominators)

Directions: Add or subtract the mixed numbers.

73. $2\frac{5}{6} + 6\frac{9}{12} = \mathbf{9\frac{7}{12}}$

$2\frac{5}{6} = \frac{17}{6} = \frac{34}{12}$
$+6\frac{9}{12} = \frac{81}{12} = \frac{81}{12}$
$\frac{115}{12} = 9\frac{7}{12}$

74.
8
$-5\frac{4}{9}$
$\overline{2\frac{5}{9}}$

$8 = \frac{8}{1} = \frac{72}{9}$
$-5\frac{4}{9} = \frac{49}{9} = \frac{49}{9}$
$\overline{\frac{23}{9} = 2\frac{5}{9}}$

75.
$6\frac{1}{4}$
$-5\frac{2}{6}$
$\overline{\frac{11}{12}}$

$6\frac{1}{4} = \frac{25}{4} = \frac{75}{12}$
$-5\frac{2}{6} = \frac{32}{6} = \frac{64}{12}$
$\overline{\frac{11}{12}}$

Directions: Multiply the fractions and reduce to lowest terms.

76. $\frac{4}{5} \times \frac{3}{6} = \mathbf{\frac{12}{30} = \frac{2}{5}}$

77. $\left(\frac{1}{6}\right)\left(\frac{9}{3}\right) = \mathbf{\frac{9}{18} = \frac{1}{2}}$

78. $-\dfrac{7}{9} \times \dfrac{4}{5} = -\dfrac{28}{45}$

79. $1\dfrac{3}{6} \times 2 = \dfrac{18}{6} = 3$ (Solution: $\dfrac{9}{6} \times \dfrac{2}{1} = \dfrac{18}{6} = 3$)

Directions: Cancel out the fractions before multiplying.

80. $\dfrac{3}{9} \times \dfrac{5}{14} \times \dfrac{7}{4} = \dfrac{5}{24}$ (Solution: $\dfrac{1}{3} \times \dfrac{5}{2} \times \dfrac{1}{4} = \dfrac{5}{24}$)

Directions: Record the fraction form and the reciprocal in the box below.

	Number	In Fraction Form	The Reciprocal
81.	$\dfrac{2}{5}$	$\dfrac{2}{5}$	$\dfrac{5}{2}$
82.	4	$\dfrac{4}{1}$	$\dfrac{1}{4}$
83.	$2\dfrac{1}{5}$	$\dfrac{11}{5}$	$\dfrac{5}{11}$

84. If you wanted to find the reciprocal of ¾, what would you multiply ¾ by?

   $\dfrac{4}{3}$

   What would the product be? **1**

   $\dfrac{3}{4} \times \dfrac{4}{3} = \dfrac{12}{12} = 1$

Directions: Divide the fractions. Reduce to lowest terms.

85. $\dfrac{3}{10} \div \dfrac{3}{5} = \dfrac{1}{2}$ (Solution: $\dfrac{3}{10} \times \dfrac{5}{3} \quad \dfrac{1}{2} \times \dfrac{1}{1} = \dfrac{1}{2}$)

86. $2 \div 3\dfrac{1}{2} = \dfrac{4}{7}$ (Solution: $\dfrac{2}{1} \div \dfrac{7}{2} \quad \dfrac{2}{1} \times \dfrac{2}{7} = \dfrac{4}{7}$)

Directions: Graph the fractions. Remember to insert the markers an equal distance apart.

87. Graph $\frac{1}{3}$.

88. Graph $\frac{-5}{3}$. (Solution: $\frac{-5}{3} = -1\frac{2}{3}$)

89. Graph $\frac{5}{3}$. (Solution: $\frac{5}{3} = 1\frac{2}{3}$)

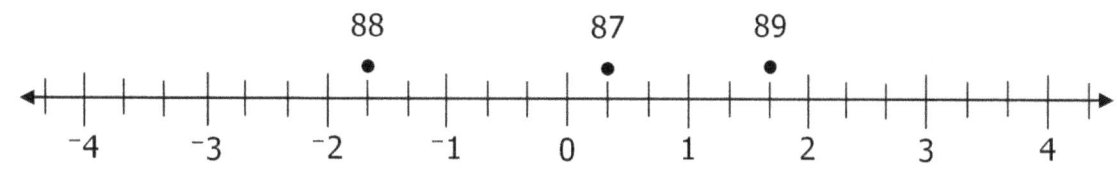

119

Name: _____   Date: _____

Key for Elementary Math Test #5
(Grades: 3rd, 4th, 5th)

Supplies: Calculator

Directions: Show all your work in the boxes provided.

Reading Numbers
Directions: Read each number and record it in the space provided.

1. 4 hundred & 21   **421**

2. 57 million, 9 hundred & 31 thousand, 4 hundred & 22   **57,931,422**

3. 6 thousand, 3 hundred & 48   **6,348**

4. seven   **7**

5. 2 billion, 3 hundred & 15 million, 5 hundred & 76 thousand, 1 hundred & 49
   2,315,576,149

6. 74 thousand, 9 hundred & 81   **74, 981**

7. 2 hundred & 54 thousand, 7 hundred & 89   **254,789**

8. 4 hundred & 51 million, 3 hundred & 75 thousand, 2 hundred & 60
   451,375,260

9. forty-three   **43**

10. 7 million, 5 hundred & 83 thousand, 4 hundred & 7   **7,583,407**

Directions: Multiply 0.41 by $9 million.

11. (0.41)($9 million) = **$3.69 million** or **$3,690,000**

$$\begin{array}{r} .41 \\ \times\ 9 \\ \hline 3.69 \end{array}$$

Rounding Numbers and Decimals

Round 394,572 to the nearest
12. 10   **394,570**
13. 100   **394,600**
14. 1,000   **395,000**
15. 10,000   **390,000**
16. 100,000   **400,000**
Round 3.8556 to the nearest
17. whole number   **4**
18. tenth   **3.9**
19. hundredth   **3.86**
20. thousandth   **3.856**

21. Round 672.597 to the nearest hundredth. **672.60**

Directions: Round the mixed number to the nearest whole number.

22. $5\frac{3}{4}$ = **6**

Directions: Divide the numbers and round the decimal quotient to the nearest hundredth. <u>Do not use a calculator.</u>

23. 11 ÷ 8 = **1.38**	$$\begin{array}{r} 1.375 \\ 8\overline{)11.000} \\ \underline{8.} \\ 3.0 \\ \underline{24} \\ 60 \\ \underline{56} \\ 40 \end{array}$$

Directions: <u>Estimate</u> by rounding the decimals to the nearest whole number. Choose a slightly larger or smaller number for the dividend (the first number) if necessary so that there is no remainder in your estimation.

24. 11.7 ÷ 4.3 ≈ **3**   (Solution: 12 ÷ 4 = 3)

25. 21.34 ÷ 9.7 ≈ **2**   (Solution: 20 ÷ 10 = 2)

The Metric System

Directions: List the below prefixes in order from the greatest value to the least.
centi- (cm), deci- (dm), deka- (dkm), hecto- (hm), kilo- (km), milli- (mm)

26. *kilo-* (km)	27. *hecto-* (hm)	28. *deka-* (dkm)	METER (m)	29. *deci-* (dm)	30. *centi-* (cm)	31. *milli-* (mm)

Directions: Convert the units of measurement.

32. 5.4 kilometers = **5,400** meters

33. 723 millimeters = **0.723** meters

Directions: Choose a number or number name from the boxes below for 34-39.

1,000	100	10	one tenth (0.1)	one hundredth (0.01)	one thousandth (0.001)

34. hectometer = **100** meters

35. decameter = **10** meters

36. centimeter = **one hundredth (0.01)** of a meter

37. millimeter = **one thousandth (0.001)** of a meter

38. kilometer = **1,000** meters

39. decimeter = **one tenth (0.1)** of a meter

Approximate Conversions
Directions: Solve the conversion problems.

40. 4 inches ≈ **100** millimeters (Solution: 4 × 25 = 100)

41. 4 inches ≈ **10** centimeters (Solution: 4 × 2.5 = 10)

42. 4 inches ≈ **0.1** meters (Solution: 4 × 0.025 = 0.1)

43. 5 feet ≈ **150** centimeters (Solution: 5 × 30 = 150)

44. 5 feet ≈ **1.5** meters (Solution: 5 × 0.3 = 1.5)

45. 6 yards ≈ **540** centimeters (Solution: 6 × 90 = 540)

46. 6 yards ≈ **5.4** meters (Solution: 6 × 0.9 = **5.4**)

Measurements (Liquids)

47. 1 tablespoon = **3** teaspoons (tsp.)

48. 1 ounce = **2** tablespoons (Tbsp.)

49. 1 cup = **8** ounces (oz.)

50. 1 pint = **2** cups

51. 1 quart = **4** cups   or   **2** pints (pt)

52. 1 gallon = **4** quarts (qt)

Conversion Problems

Directions: Solve the conversion problems.

53.	2 tablespoons = **6** teaspoons (Solution: 3 tsp = 1 Tbsp; 2 × 3 = 6)
54.	9 tsp = **3** Tbsp (Solution: 3 tsp = 1 Tbsp; 9 ÷ 3 = 3)
55.	4 tablespoons = **2** ounces (Solution: 2 Tbsp = 1 oz; 4 ÷ 2 = 2)
56.	3 oz = **4** Tbsp (Solution: 2 Tbsp = 1 oz; 3 × 2 = 6)
57.	2 cups = **16** ounces (Solution: 8 oz = 1 cup; 2 × 8 = 16)
58.	24 ounces = **3** cups (Solution: 8 oz = 1 cup; 24 ÷ 8 = 3)
59.	4 cups = **2** pints (Solution: 2 cups = 1 pt; 4 ÷ 2 = 2)
60.	3 pints = **6** cups (Solution: 2 cups = 1 pt; 3 × 2 = 6)
61.	2 quarts = **4** pints (Solution: 2 pints = 1 quart; 2 × 2 = 4)
62.	6 pints = **3** quarts (Solution: 2 pints = 1 quart; 6 ÷ 2 = 3)

63. 12 quarts = **3** gallons (Solution: 4 qt = 1 gal; 12 ÷ 4 = 3)

64. 4 gal = **16** qt (Solution: 4 qt = 1 gal; 4 × 4 = 16)

Measurements (Weight)

65. Describe the weight of a gram (g) and tell how many milligrams are in one gram.

   One raisin weighs about one gram. A gram weighs even less than one ounce. There are 1,000 milligrams in one gram.

66. 16 ounces = **1 pound**

67. kilogram (kg) = **1,000 grams (less than 3 pounds)**

Conversion Problems

Directions: Solve the conversion problems.

68. 2 pounds = **32** ounces (Solution: 16 oz = 1 pound; 2 × 16 = 32)

69. 48 ounces = **3** pounds (Solution: 16 oz = 1 pound; 48 ÷ 16 = 3)

70. 2,000 grams = **2** kilograms
     (Solution: 1 kg = 1000 grams; 2000 ÷ 1000 = 2)

71. 3 kilograms = **3,000** grams
     (Solution: 1 kg = 1000 grams; 3 × 1000 = 3000)

Directions: Fill in the blanks.

72. According to the Fahrenheit temperature scale, water freezes at **32°**.

73. According to the Fahrenheit temperature scale, water boils at **212°**.

74. According to the Fahrenheit temperature scale, normal is **98.6°**.

75. The formula used to convert Fahrenheit into Celsius is **(F − 32) × 5 ÷ 9**.

76. Use your calculator to convert 212° Fahrenheit into Celsius.
    100° C (Solution: 212 − 32 = 180, 180 × 5 ÷ 9 = 100°)

Directions: Fill in the blanks.

77. According to the Celsius temperature scale, water freezes at **0°**.

78. According to the Celsius temperature scale, water boils at **100°**.

79. According to the Celsius temperature scale, normal is **37°**.

80. The formula used to convert Celsius into Fahrenheit is **C × 9 ÷ 5 + 32**.

81. Convert 100° Celsius into Fahrenheit.
    212° F (Solution: 100 × 9 ÷ 5 + 32 = 212°)

Least Common Multiple (LCM)

Directions: Identify the first nine nonzero multiples of the numbers 3 and 7. Then find the least common multiple (LCM) of 3 and 7.

82. Multiples of 3: **3, 6, 9, 12, 15, 18, 21, 24, 27**

83. Multiples of 7: **7, 14, 21, 28, 35, 42, 49, 56, 63**

84. What is the LCM of 3 and 7? **21**

Greatest Common Factor (GCF)

> 85. If you are looking for the greatest common factor (GCF) of two numbers, such as 6 and 12, you are looking for what?
>
>   **The largest number that both numbers can be divided evenly by**
>
> 86. What are the factors of 10? **1, 2, 5, 10**
>
> 87. What are the factors of 20? **1, 2, 4, 5, 10, 20**
>
> 88. What are the common factors of 10 and 20? **1, 2, 5, 10**
>
> 89. What is the GCF of 10 and 20? **10**
>
> 90. Are the numbers 10 and 20 **relatively prime**? Tell how you know.
>
>   **No because their GCF is not 1**

Can the following numbers be divided evenly by 2, by 5, by 10, or by all three of the numbers?

91. The number 75 is divisible by **5**.

92. The number 54 is divisible by **2**.

93. The number 500 is divisible by **2, 5, and 10**.

Prime and Composite Numbers

94. Is the number **10** prime, composite, or neither prime nor composite?

    Composite

95. Is the number **13** prime, composite, or neither prime nor composite?

    Prime

96. Is the number **1** prime, composite, or neither prime nor composite?

    Neither

True or False
97. A number less than 2 (such as −2, −1, 0, or 1) cannot be prime or composite.

    True

98. Use factor trees to find the prime factorization of number 18. Use exponents whenever possible. **2 × 3² or 3² × 2**

```
      18                    18
     /  \                  /  \
    2 × 9                 3 × 6
       /  \                  /  \
   2 × 3 × 3             3 × 2 × 3
```

99. What are the prime factors of 42? **7 × 2 × 3**

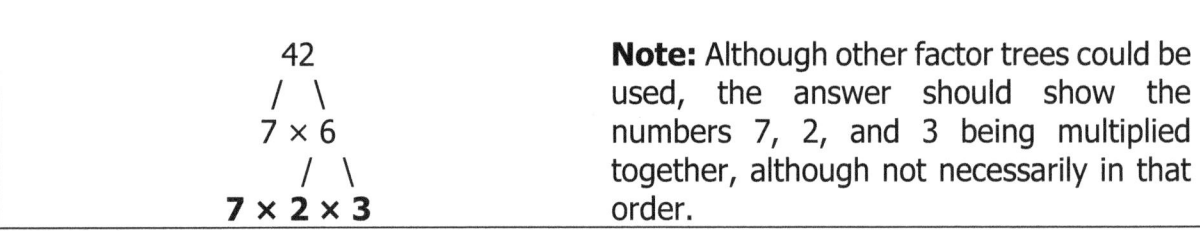

```
      42
     /  \
    7 × 6
       /  \
   7 × 2 × 3
```

Note: Although other factor trees could be used, the answer should show the numbers 7, 2, and 3 being multiplied together, although not necessarily in that order.

Directions: Use factor trees to locate the GCF and the LCM of each pair of numbers below.

100. GCF of 10 and 15: **5**

101. LCM of 10 and 15: **30** (Solution: 5 × 2 × 3 = 30)

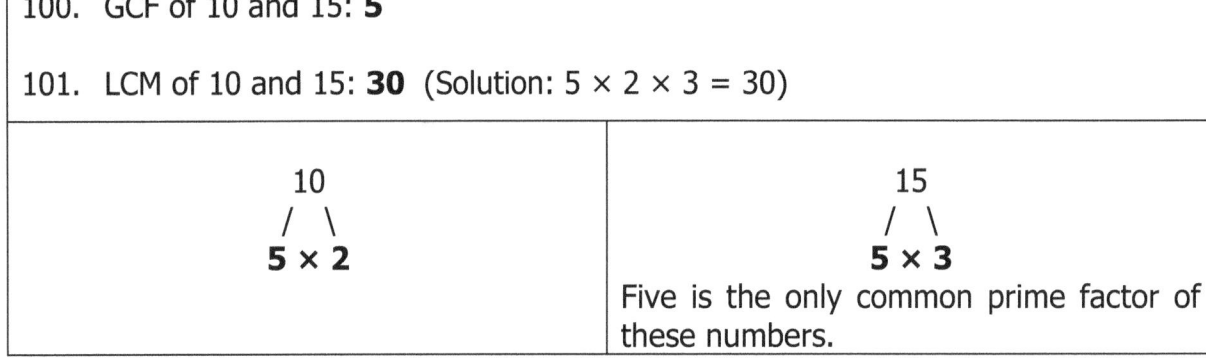

```
      10                    15
     /  \                  /  \
    5 × 2                 5 × 3
```

Five is the only common prime factor of these numbers.

102. GCF of 42 and 70: **14** (Solution: 7 × 2 = 14)

103. LCM of 42 and 70: **210** (Solution: 7 × 3 × 2 × 5 = 210)

```
      42                    70
     /  \                  /  \
    7 × 6                 7 × 10
       /  \                  /  \
   7 × 3 × 2             7 × 5 × 2
```

Name: _____   Date: _____

Key for Elementary Geometry Test #6
(Grades: 3rd, 4th, 5th)

Supplies: Protractor

Directions: Show all your work in the boxes provided.

Geometry studies points, lines, and planes, and the shapes they create. Sketch these figures in the spaces provided.

1. Point	2. Line	3. Plane
•	←——→	△

Directions: Identify the lines below. Be as specific as possible.
intersecting lines, parallel lines, perpendicular lines

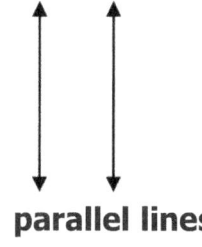   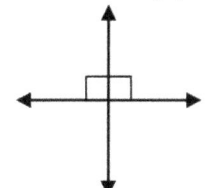   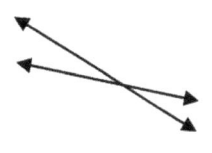   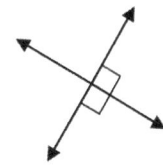

     parallel lines   **perpendicular**   **intersecting lines**   **perpendicular**
4. _____   5. _____   6. _____   7. _____

8. What do the arrowheads of a line signify?

   That the line continues endlessly in both directions

Directions: Fill in the blanks.

9. A **line** has at least two points.

10. A **plane** is flat. It has at least three points that are not all on the same line.

11. **Space** has at least four points that are not all lying in the same plane.

12. **Parallel** lines do not intersect, and they are the same distance apart.

Directions: Classify each figure in the box as a *simple curve*, a *simple closed curve*, a *closed curve*, or a *curve*.

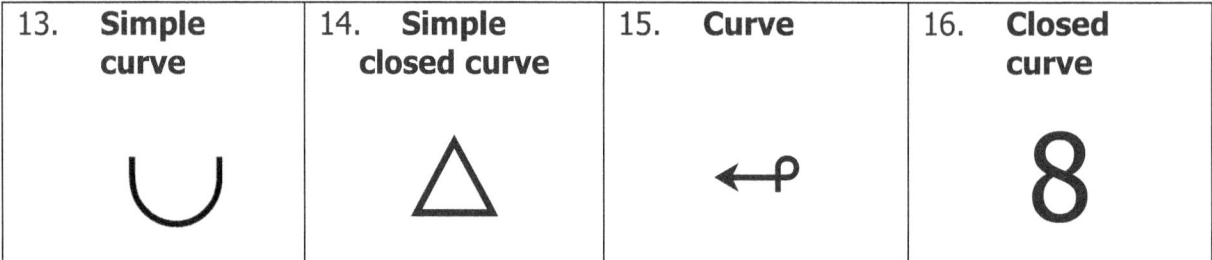

13. **Simple curve**	14. **Simple closed curve**	15. **Curve**	16. **Closed curve**

17. How do you distinguish curves from simple curves? Use the illustrations above to help you.

    A curve can cross itself, but a simple curve can only intersect itself when the starting and ending points touch, as in a simple closed curve.

Directions: Identify the two polygons below as simple or complex. Tell how you know the polygon is simple or complex.

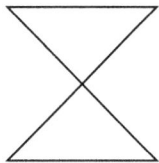   

18. **This is a complex polygon because it crosses over itself.**

19. **This is a simple polygon; it does not cross over itself.**

20. Cross out the shapes that are not polygons.

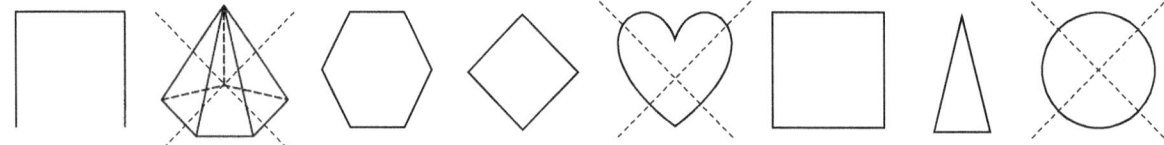

21. The first shape above is a polygonal curve, but is it also a polygon? Explain your answer.

    No because it is not a closed shape; it is an open figure.

22. Which shape above is an irregular polygon (a polygon that is not <u>regular</u>)? Explain your answer.

    The triangle is irregular because all its sides are not the same length, and all its angles are not the same measure.

23. What does *equilateral* mean?

    All the sides of a shape are the same length.

24. What does *equiangular* mean?

    All the angles of a shape have the same measure.

25. What are the two dimensions of a plane (flat) shape?

    length and height

26. What are the three dimensions of a 3-D (solid) figure?

    length, height, and depth

Directions: Record the number of sides, angles, and vertices each polygon has.
3, 4, 5, 6, 7, 8, 9, 10, 11, 12, or *unknown*

27.	nonagon	**9**	33.	heptagon	**7**
28.	dodecagon	**12**	34.	quadrilateral	**4**
29.	octagon	**8**	35.	decagon	**10**
30.	pentagon	**5**	36.	triangle	**3**
31.	hendecagon	**11**	37.	hexagon	**6**
32.	*n*-gon	**unknown**	38.	What is a 5-gon?	**pentagon**

39. A trapezoid has one pair of parallel opposite sides, called *bases*. Put an *x* through the parallel sides in the trapezoid below.

    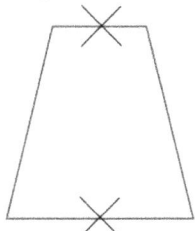

40. Fill in the blanks: A **rhombus** is a parallelogram with all four sides the same length.

    A **square** is a rectangle with all four sides the same length.

41. Circle the concave shapes below.

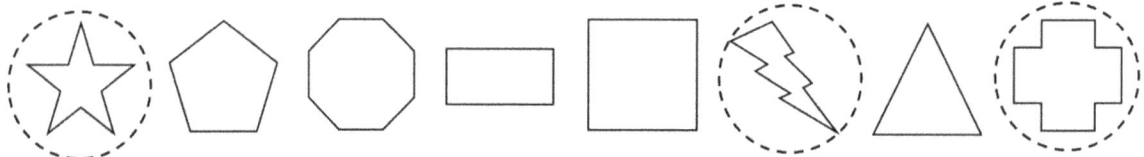

42. How do you know that the shapes you circled are concave rather than convex?

   If you were to draw a line from one point on, for example, the star to a point next to it, the line would go outside the shape.

Directions: Identify the number of faces each polyhedron has.
  4, 5, 6, 7, 8, 10

43. Octahedron **8**

44. Hexahedron **6**

45. Pentahedron **5**

46. Decahedron **10**

47. Tetrahedron **4**

48. Heptahedron **7**

Directions: Write the name of each shape in the space provided. Use a word from the list.

circle, decagon, dodecagon, hendecagon, heptagon, hexagon, nonagon, octagon, oval, parallelogram, pentagon, rectangle, rhombus, square, trapezoid, triangle

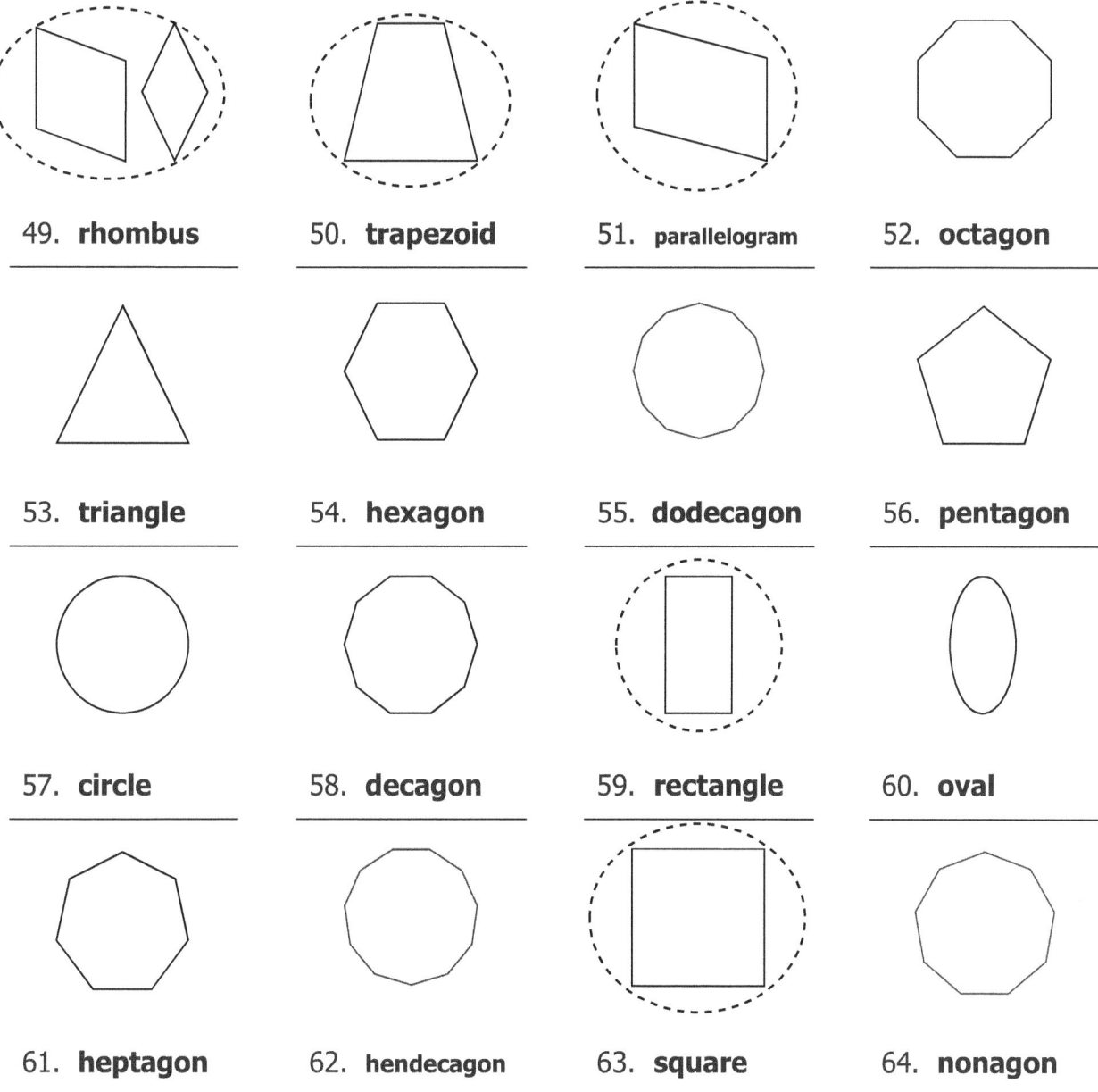

49. **rhombus**
50. **trapezoid**
51. **parallelogram**
52. **octagon**
53. **triangle**
54. **hexagon**
55. **dodecagon**
56. **pentagon**
57. **circle**
58. **decagon**
59. **rectangle**
60. **oval**
61. **heptagon**
62. **hendecagon**
63. **square**
64. **nonagon**

65. Circle all the shapes on this page that are also quadrilaterals.

Directions: Write the name of each solid figure in the space provided. Use a word from the list.

cone, cube, cylinder, hexagonal prism, hexagonal pyramid, pentagonal prism, pentagonal pyramid, rectangular prism (or cuboid), sphere, triangular prism, triangular pyramid

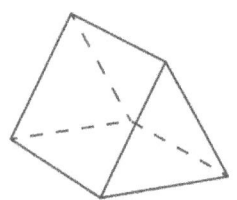
66. triangular Prism

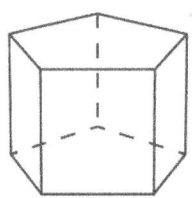
67. pentagonal prism

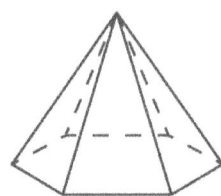
68. hexagonal pyramid

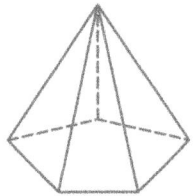
69. pentagonal pyramid

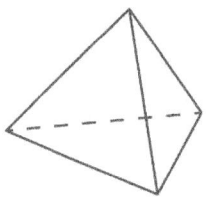
70. triangular pyramid

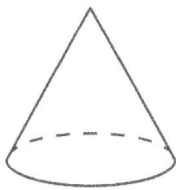
71. cone

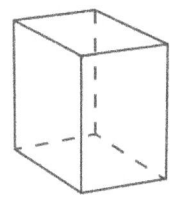
72. rectangular prism

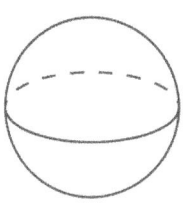
73. sphere

74. cylinder

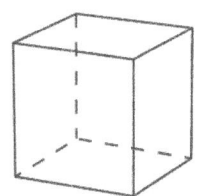
75. cube

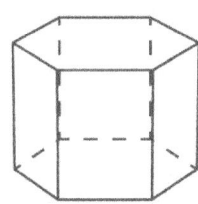
76. hexagonal prism

77. Which two shapes above are also called heptahedrons?

pentagonal prism, hexagonal pyramid

Directions: Fill in the blanks. Use each of the choices once and be as specific as possible.

pyramid, geometric solid (or three-dimensional figure), polyhedron, prism

78. A **geometric solid (or three-dimensional figure)** is not a flat shape.

79. A 3-D figure that has polygons for faces is a **polyhedron**.

80. A **prism** has two congruent polygonal faces, called bases, and all the rest of its faces are quadrilaterals.

81. A **pyramid** has only one polygonal base, and the rest of its faces are triangles.

82. Why is a cylinder not a prism or a polyhedron?

    Because the circular faces are not polygons

83. Euler's Formula, which is true for any convex polyhedron, is $V - E + F = 2$.

Directions: Identify the number of vertices, edges, and faces the figure has. Then compare Euler's Formula with the number of vertices, edges, and faces.

84. Vertices: **6**
85. Edges: **9**
86. Faces: **5**
87. Euler's Formula:
    $6 - 9 + 5 = 2$

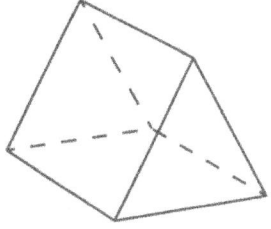

Directions: Identify the angle. Write *right, acute, obtuse, straight.,* or *zero angle.*

88. **obtuse**   89. **straight**   90. **right**   91. **acute**

Directions: Complete the sentences.

92. An **acute angle** measures **less than 90° and more than 0°**.

135

93. A **right angle** measures **90°**.

94. An **obtuse angle** measures **more than 90° but less than 180°**.

95. A **straight angle** measures **180°**.

96. I estimate the measure of this angle to be **35°**.
    A. 35°   C. 95°
    B. 85°   D. 115°

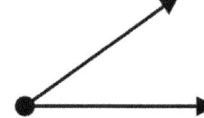

On this test, your protractor may only be used for drawing the angle.

97. In the box below, use your protractor to draw a 70° angle. Name the angle you drew ∠PQR. **(Measure each child's angle with a protractor.)**

98. In the space provided, list two other ways of naming ∠PQR.

    ∠Q and ∠RQP

99. What notation could you use to show that the angle measures 70°?

    m∠PQR = 70°

100. What is the vertex of the 70° angle you drew? **Q**

101. What are the rays of the 70° angle you drew? Use the correct notation.

    $\overrightarrow{QP}$ **and** $\overrightarrow{QR}$

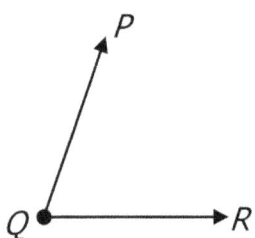

136

Directions: Find the area of the square, the rectangle, and the parallelogram.

Square	Rectangle	Parallelogram
102. **25** in² (Solution: 5 × 5 = 25) 5 in.	103. **21** in² (Solution: 3 × 7 = 21) 3 in. 7 in.	104. **15** ft² (Solution: 5 × 3 = 15) 5 ft. 3 ft.

105. Fill in the blanks: A radius is ½ the length of the diameter.

   (A radius goes **halfway** through a circle.)

106. If the diameter of a circle is 14 inches, what is its radius?

   7 inches

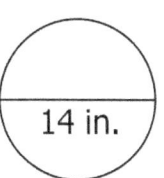

107. This circle has a radius of 2 cm. What is its diameter?

   4

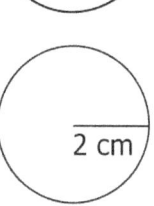

108. Do reflections (flips), rotations (turns), and translations (slides) change in position, size, or shape?

   position

Directions: Identify the type of transformation below—*reflection (flip), rotation (turn),* or *translation (slide)*.

109. **Rotation (turn)**	110. **Translation (slide)**	111. **Reflection (flip)**

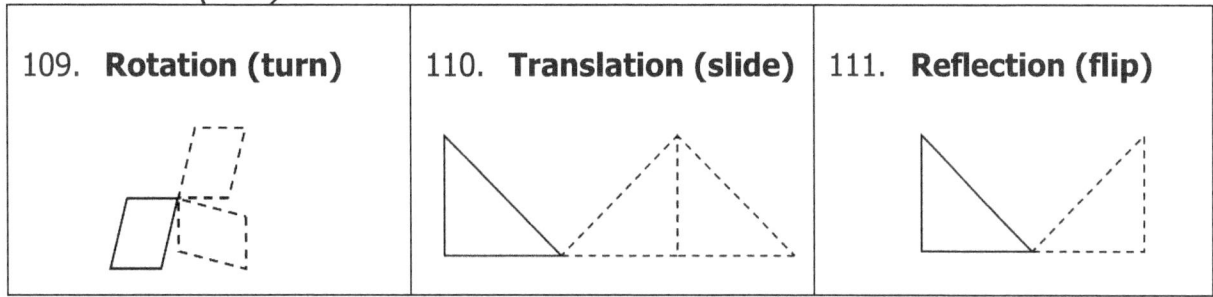

www.ingramcontent.com/pod-product-compliance
Lightning Source LLC
Chambersburg PA
CBHW081324040426
42453CB00013B/2292